www.EffortlessMath.com

... So Much More Online!

✓ FREE Math lessons

✓ More Math learning books!

✓ Mathematics Worksheets

✓ Online Math Tutors

Need a PDF version of this book?

Please visit www.EffortlessMath.com

ISEE Middle Level Math Prep 2020-2021

The Most Comprehensive Review and Ultimate Guide to the ISEE Middle Level Math Test

By

Reza Nazari & Ava Ross

ISBN: 978-1-64612-308-7

Published by: Effortless Math Education

www.EffortlessMath.com

Visit www.EffortlessMath.com
for Online Math Practice

Description

ISEE Middle Level Math Prep 2020 – 2021, which reflects the 2020 - 2021 test guidelines, is dedicated to preparing test takers to ace the ISEE Middle Level Math Test. This comprehensive ISEE Middle Level Math Prep book with hundreds of examples, abundant sample ISEE Middle Level mathematics questions, and two full-length and realistic ISEE Middle Level Math tests is all you will ever need to fully prepare for the ISEE Middle Level Math. It will help you learn everything you need to ace the math section of the ISEE Middle Level test.

Effortless Math unique study program provides you with an in-depth focus on the math portion of the exam, helping you master the math skills that students find the most troublesome. This ISEE Middle Level Math preparation book contains most common sample questions that are most likely to appear in the mathematics section of the ISEE Middle Level.

Inside the pages of this comprehensive ISEE Middle Level Math book, students can learn basic math operations in a structured manner with a complete study program to help them understand essential math skills. It also has many exciting features, including:

- ✓ Content 100% aligned with the 2020 ISEE Middle Level test
- ✓ Written by ISEE Middle Level Math instructors and test experts
- ✓ Complete coverage of all ISEE Middle Level Math concepts and topics which you will be tested
- ✓ Over 2,500 additional ISEE Middle Level math practice questions in both multiple-choice and grid-in formats with answers grouped by topic, so you can focus on your weak areas
- ✓ Abundant Math skill building exercises to help test-takers approach different question types that might be unfamiliar to them
- ✓ Exercises on different ISEE Middle Level Math topics such as integers, percent, equations, polynomials, exponents and radicals
- ✓ 2 full-length practice tests (featuring new question types) with detailed answers

ISEE Middle Level Math Prep 2020 – 2021 is an incredibly useful resource for those who want to review all topics being covered on the ISEE Middle Level test. It efficiently and effectively reinforces learning outcomes through engaging questions and repeated practice, helping you to quickly master Math skills.

About the Author

Reza Nazari is the author of more than 100 Math learning books including:
– **Math and Critical Thinking Challenges:** For the Middle and High School Student
– **ACT Math in 30 Days**
– **ASVAB Math Workbook 2018 - 2019**
– **Effortless Math Education Workbooks**
– **and many more Mathematics books …**

Reza is also an experienced Math instructor and a test–prep expert who has been tutoring students since 2008. Reza is the founder of Effortless Math Education, a tutoring company that has helped many students raise their standardized test scores—and attend the colleges of their dreams. Reza provides an individualized custom learning plan and the personalized attention that makes a difference in how students view math.

You can contact Reza via email at:
reza@EffortlessMath.com

Find Reza's professional profile at:
goo.gl/zoC9rJ

Contents

Chapter 1:

Fractions and Mixed Numbers

Math Topics that you'll learn in this Chapter:

- ✓ Simplifying Fractions

- ✓ Adding and Subtracting Fractions

- ✓ Multiplying and Dividing Fractions

- ✓ Adding Mixed Numbers

- ✓ Subtracting Mixed Numbers

- ✓ Multiplying Mixed Numbers

- ✓ Dividing Mixed Numbers

Simplifying Fractions

☑ A fraction contains two numbers separated by a bar in between them. The bottom number, called the denominator, is the total number of equally divided portions in one whole. The top number, called the numerator, is how many portions you have. And the bar represents the operation of division.

☑ Simplifying a fraction means reducing it to lowest terms. To simplify a fraction, evenly divide both the top and bottom of the fraction by $2, 3, 5, 7, \ldots$ etc.

☑ Continue until you can't go any further.

Examples:

1) Simplify $\frac{12}{30}$

Solution: To simplify $\frac{12}{30}$, find a number that both 12 and 30 are divisible by.

Both are divisible by 6. Then: $\frac{12}{30} = \frac{12 \div 6}{30 \div 6} = \frac{2}{5}$

2) Simplify $\frac{64}{80}$

Solution: To simplify $\frac{64}{80}$, find a number that both 64 and 80 are divisible by.

Both are divisible by 8 and 16. Then: $\frac{64}{80} = \frac{64 \div 8}{80 \div 8} = \frac{8}{10}$, 8 and 10 are divisible by 2,

then: $\frac{8}{10} = \frac{4}{5}$ or $\frac{64}{80} = \frac{64 \div 16}{80 \div 16} = \frac{4}{5}$

3) Simplify $\frac{20}{60}$

Solution: To simplify $\frac{20}{60}$, find a number that both 20 and 60 are divisible by. Both are divisible by 20.

Then: $\frac{20}{60} = \frac{20 \div 20}{60 \div 20} = \frac{1}{3}$

Adding and Subtracting Fractions

☑ For "like" fractions (fractions with the same denominator), add or subtract the numerators (top numbers) and write the answer over the common denominator (bottom numbers).

☑ Adding and Subtracting fractions with the same denominator:

$$\frac{a}{b} + \frac{c}{b} = \frac{a+c}{b} \qquad\qquad \frac{a}{b} - \frac{c}{b} = \frac{a-c}{b}$$

☑ Find equivalent fractions with the same denominator before you can add or subtract fractions with different denominators.

☑ Adding and Subtracting fractions with different denominators:

$$\frac{a}{b} + \frac{c}{d} = \frac{ad+bc}{bd} \qquad\qquad \frac{a}{b} - \frac{c}{d} = \frac{ad-bc}{bd}$$

Examples:

1) Find the sum. $\frac{3}{4} + \frac{1}{3} =$

Solution: These two fractions are "unlike" fractions. (they have different denominators). Use this formula: $\frac{a}{b} + \frac{c}{d} = \frac{ad+cb}{bd}$

Then: $\frac{3}{4} + \frac{1}{3} = \frac{(3)(3)+(4)(1)}{4 \times 3} = \frac{9+4}{12} = \frac{13}{12}$

2) Find the difference. $\frac{4}{5} - \frac{3}{7} =$

Solution: For "unlike" fractions, find equivalent fractions with the same denominator before you can add or subtract fractions with different denominators. Use this formula:

$\frac{a}{b} - \frac{c}{d} = \frac{ad-bc}{bd}$

$\frac{4}{5} - \frac{3}{7} = \frac{(4)(7)-(3)(5)}{5 \times 7} = \frac{28-15}{35} = \frac{13}{35}$

Multiplying and Dividing Fractions

☑ Multiplying fractions: multiply the top numbers and multiply the bottom numbers. Simplify if necessary. $\frac{a}{b} \times \frac{c}{d} = \frac{a \times c}{b \times d}$

☑ Dividing fractions: Keep, Change, Flip

Keep first fraction, change division sign to multiplication, and flip the numerator and denominator of the second fraction. Then, solve! $\frac{a}{b} \div \frac{c}{d} = \frac{a}{b} \times \frac{d}{c} = \frac{a \times d}{b \times c}$

Examples:

1) Multiply. $\frac{5}{8} \times \frac{2}{3} =$

Solution: Multiply the top numbers and multiply the bottom numbers.
$\frac{5}{8} \times \frac{2}{3} = \frac{5 \times 2}{8 \times 3} = \frac{10}{24}$, simplify: $\frac{10}{24} = \frac{10 \div 2}{24 \div 2} = \frac{5}{12}$

2) Solve. $\frac{1}{3} \div \frac{4}{7} =$

Solution: Keep first fraction, change division sign to multiplication, and flip the numerator and denominator of the second fraction.
Then: $\frac{1}{3} \div \frac{4}{7} = \frac{1}{3} \times \frac{7}{4} = \frac{1 \times 7}{3 \times 4} = \frac{7}{12}$

3) Calculate. $\frac{3}{5} \times \frac{2}{3} =$

Solution: Multiply the top numbers and multiply the bottom numbers.
$\frac{3}{5} \times \frac{2}{3} = \frac{3 \times 2}{5 \times 3} = \frac{6}{15}$, simplify: $\frac{6}{15} = \frac{6 \div 3}{15 \div 3} = \frac{2}{5}$

4) Solve. $\frac{1}{4} \div \frac{5}{6} =$

Solution: Keep first fraction, change division sign to multiplication, and flip the numerator and denominator of the second fraction.
Then: $\frac{1}{4} \div \frac{5}{6} = \frac{1}{4} \times \frac{6}{5} = \frac{1 \times 6}{4 \times 5} = \frac{6}{20}$, simplify: $\frac{6}{20} = \frac{6 \div 2}{20 \div 2} = \frac{3}{10}$

Adding Mixed Numbers

Use following steps for adding mixed numbers:

☑ Add whole numbers of the mixed numbers.

☑ Add the fractions of the mixed numbers.

☑ Find the Least Common Denominator (LCD) if necessary.

☑ Add whole numbers and fractions.

☑ Write your answer in lowest terms.

Examples:

1) Add mixed numbers. $3\frac{1}{3} + 1\frac{4}{5} =$

Solution: Let's rewriting our equation with parts separated, $3\frac{1}{3} + 1\frac{4}{5} = 3 + \frac{1}{3} + 1 + \frac{4}{5}$. Now, add whole number parts: $3 + 1 = 4$

Add the fraction parts $\frac{1}{3} + \frac{4}{5}$. Rewrite to solve with the equivalent fractions. $\frac{1}{3} + \frac{4}{5} = \frac{5}{15} + \frac{12}{15} = \frac{17}{15}$. The answer is an improper fraction (numerator is bigger than denominator). Convert the improper fraction into a mixed number: $\frac{17}{15} = 1\frac{2}{15}$. Now, combine the whole and fraction parts: $4 + 1\frac{2}{15} = 5\frac{2}{15}$

2) Find the sum. $1\frac{2}{5} + 2\frac{1}{2} =$

Solution: Rewriting our equation with parts separated, $1 + \frac{2}{5} + 2 + \frac{1}{2}$. Add the whole number parts:

$1 + 2 = 3$. Add the fraction parts: $\frac{2}{5} + \frac{1}{2} = \frac{4}{10} + \frac{5}{10} = \frac{9}{10}$

Now, combine the whole and fraction parts: $3 + \frac{9}{10} = 3\frac{9}{10}$

ISEE Middle Level Math Prep 2020-2021

Subtract Mixed Numbers

Use the following steps for subtracting mixed numbers.

☑ Convert mixed numbers into improper fractions. $a\frac{c}{b} = \frac{ab+c}{b}$

☑ Find equivalent fractions with the same denominator for unlike fractions. (fractions with different denominators)

☑ Subtract the second fraction from the first one. $\frac{a}{b} - \frac{c}{d} = \frac{ad-bc}{bd}$

☑ Write your answer in lowest terms.

☑ If the answer is an improper fraction, convert it into a mixed number.

Examples:

1) Subtract. $3\frac{4}{5} - 1\frac{3}{4} =$

   Solution: Convert mixed numbers into fractions: $3\frac{4}{5} = \frac{3\times5+4}{5} = \frac{19}{5}$ and $1\frac{3}{4} = \frac{1\times4+3}{4} = \frac{7}{4}$
   These two fractions are "unlike" fractions. (they have different denominators). Find equivalent fractions with the same denominator. Use this formula: $\frac{a}{b} - \frac{c}{d} = \frac{ad-bc}{bd}$
   $\frac{19}{5} - \frac{7}{4} = \frac{(19)(4)-(5)(7)}{5\times4} = \frac{76-35}{20} = \frac{41}{20}$, the answer is an improper fraction, convert it into a mixed number. $\frac{41}{20} = 2\frac{1}{20}$

2) Subtract. $4\frac{3}{8} - 1\frac{1}{2} =$

   Solution: Convert mixed numbers into fractions: $4\frac{3}{8} = \frac{4\times8+3}{8} = \frac{35}{8}$ and $1\frac{1}{2} = \frac{1\times2+1}{4} = \frac{3}{2}$
   Find equivalent fractions: $\frac{3}{2} = \frac{12}{8}$. Then: $4\frac{3}{8} - 1\frac{1}{2} = \frac{35}{8} - \frac{12}{8} = \frac{23}{8}$
   The answer is an improper fraction, convert it into a mixed number.
   $$\frac{23}{8} = 2\frac{7}{8}$$

Multiplying Mixed Numbers

Use following steps for multiplying mixed numbers:

☑ Convert the mixed numbers into fractions. $a\frac{c}{b} = a + \frac{c}{b} = \frac{ab+c}{b}$

☑ Multiply fractions. $\frac{a}{b} \times \frac{c}{d} = \frac{a \times c}{b \times d}$

☑ Write your answer in lowest terms.

☑ If the answer is an improper fraction (numerator is bigger than denominator), convert it into a mixed number.

Examples:

1) Multiply. $3\frac{1}{3} \times 4\frac{1}{6} =$

   Solution: Convert mixed numbers into fractions, $3\frac{1}{3} = \frac{3 \times 3 + 1}{3} = \frac{10}{3}$ and $4\frac{1}{6} = \frac{4 \times 6 + 1}{6} = \frac{25}{6}$

   Apply the fractions rule for multiplication, $\frac{10}{3} \times \frac{25}{6} = \frac{10 \times 25}{3 \times 6} = \frac{250}{18}$

   The answer is an improper fraction. Convert it into a mixed number. $\frac{250}{18} = 13\frac{8}{9}$

2) Multiply. $2\frac{1}{2} \times 3\frac{2}{3} =$

   Solution: Converting mixed numbers into fractions, $2\frac{1}{2} \times 3\frac{2}{3} = \frac{5}{2} \times \frac{11}{3}$

   Apply the fractions rule for multiplication, $\frac{5}{2} \times \frac{11}{3} = \frac{5 \times 11}{2 \times 3} = \frac{55}{6} = 9\frac{1}{6}$

3) Multiply mixed numbers. $2\frac{1}{3} \times 2\frac{1}{2} =$

   Solution: Converting mixed numbers to fractions, $2\frac{1}{3} = \frac{7}{3}$ and $2\frac{1}{2} = \frac{5}{2}$. Multiply two fractions:

   $$\frac{7}{3} \times \frac{5}{2} = \frac{7 \times 5}{3 \times 2} = \frac{35}{6} = 5\frac{5}{6}$$

Dividing Mixed Numbers

Use following steps for dividing mixed numbers:

☑ Convert the mixed numbers into fractions. $a\frac{c}{b} = a + \frac{c}{b} = \frac{ab+c}{b}$

☑ Divide fractions: Keep, Change, Flip: Keep first fraction, change division sign to multiplication, and flip the numerator and denominator of the second fraction. Then, solve! $\frac{a}{b} \div \frac{c}{d} = \frac{a}{b} \times \frac{d}{c} = \frac{a \times d}{b \times c}$

☑ Write your answer in lowest terms.

☑ If the answer is an improper fraction (numerator is bigger than denominator), convert it into a mixed number.

Examples:

1) Solve. $3\frac{2}{3} \div 2\frac{1}{2}$

Solution: Convert mixed numbers into fractions: $3\frac{2}{3} = \frac{3\times3+2}{3} = \frac{11}{3}$ and $2\frac{1}{2} = \frac{2\times2+1}{2} = \frac{5}{2}$

Keep, Change, Flip: $\frac{11}{3} \div \frac{5}{2} = \frac{11}{3} \times \frac{2}{5} = \frac{11\times2}{3\times5} = \frac{22}{15}$. The answer is an improper fraction. Convert it into a mixed number: $\frac{22}{15} = 1\frac{7}{15}$

2) Solve. $3\frac{4}{5} \div 1\frac{5}{6}$

Solution: Convert mixed numbers to fractions, then solve:

$$3\frac{4}{5} \div 1\frac{5}{6} = \frac{19}{5} \div \frac{11}{6} = \frac{19}{5} \times \frac{6}{11} = \frac{114}{55} = 2\frac{4}{55}$$

3) Solve. $2\frac{2}{7} \div 2\frac{3}{5}$

Solution: Converting mixed numbers to fractions: $3\frac{4}{5} \div 1\frac{5}{6} = \frac{16}{7} \div \frac{13}{5}$

Keep, Change, Flip: $\frac{16}{7} \div \frac{13}{5} = \frac{16}{7} \times \frac{5}{13} = \frac{16\times5}{7\times13} = \frac{80}{91}$

Chapter 1: Practices

✍ *Simplify each fraction.*

1) $\dfrac{18}{30} =$ 3) $\dfrac{35}{55} =$ 5) $\dfrac{54}{81} =$

2) $\dfrac{21}{42} =$ 4) $\dfrac{48}{72} =$ 6) $\dfrac{80}{200} =$

✍ *Find the sum or difference.*

7) $\dfrac{6}{15} + \dfrac{3}{15} =$ 9) $\dfrac{1}{4} + \dfrac{2}{5} =$ 11) $\dfrac{1}{2} - \dfrac{3}{8} =$

8) $\dfrac{2}{3} + \dfrac{1}{9} =$ 10) $\dfrac{7}{10} - \dfrac{3}{10} =$ 12) $\dfrac{5}{7} - \dfrac{3}{5} =$

✍ *Find the answers.*

13) $\dfrac{1}{7} \div \dfrac{3}{8} =$ 15) $\dfrac{5}{7} \times \dfrac{3}{4} =$ 17) $\dfrac{3}{7} \div \dfrac{5}{8} =$

14) $\dfrac{2}{3} \times \dfrac{4}{7} =$ 16) $\dfrac{2}{5} \div \dfrac{3}{7} =$ 18) $\dfrac{3}{8} \times \dfrac{4}{7} =$

✍ *Calculate.*

19) $3\dfrac{1}{5} + 2\dfrac{2}{9} =$ 21) $4\dfrac{4}{5} + 1\dfrac{2}{7} =$ 23) $1\dfrac{5}{6} + 1\dfrac{2}{5} =$

20) $1\dfrac{1}{7} + 5\dfrac{2}{5} =$ 22) $2\dfrac{4}{7} + 2\dfrac{3}{5} =$ 24) $3\dfrac{5}{7} + 1\dfrac{2}{9} =$

✎ **Calculate.**

25) $3\frac{2}{5} - 1\frac{2}{9} =$

26) $5\frac{3}{5} - 1\frac{1}{7} =$

27) $4\frac{2}{5} - 2\frac{2}{7} =$

28) $8\frac{3}{4} - 2\frac{1}{8} =$

29) $9\frac{5}{7} - 7\frac{4}{21} =$

30) $11\frac{7}{12} - 9\frac{5}{6} =$

✎ **Find the answers.**

31) $1\frac{1}{8} \times 1\frac{3}{4} =$

32) $3\frac{1}{5} \times 2\frac{2}{7} =$

33) $2\frac{1}{8} \times 1\frac{2}{9} =$

34) $2\frac{3}{8} \times 2\frac{2}{5} =$

35) $1\frac{1}{2} \times 5\frac{2}{3} =$

36) $3\frac{1}{2} \times 6\frac{2}{3} =$

✎ **Solve.**

37) $9\frac{1}{2} \div 2\frac{3}{5} =$

38) $2\frac{3}{8} \div 1\frac{2}{5} =$

39) $5\frac{3}{4} \div 2\frac{2}{7} =$

40) $8\frac{1}{3} \div 4\frac{1}{4} =$

41) $7\frac{2}{5} \div 3\frac{3}{4} =$

42) $2\frac{4}{5} \div 3\frac{2}{3} =$

Answers – Chapter 1

1) $\frac{3}{5}$

2) $\frac{1}{2}$

3) $\frac{7}{11}$

4) $\frac{2}{3}$

5) $\frac{2}{3}$

6) $\frac{2}{5}$

7) $\frac{3}{5}$

8) $\frac{7}{9}$

9) $\frac{13}{20}$

10) $\frac{2}{5}$

11) $\frac{1}{8}$

12) $\frac{4}{35}$

13) $\frac{8}{21}$

14) $\frac{8}{21}$

15) $\frac{15}{28}$

16) $\frac{14}{15}$

17) $\frac{24}{35}$

18) $\frac{3}{14}$

19) $5\frac{19}{45}$

20) $6\frac{19}{35}$

21) $6\frac{3}{35}$

22) $5\frac{6}{35}$

23) $3\frac{7}{30}$

24) $4\frac{59}{63}$

25) $2\frac{8}{45}$

26) $6\frac{16}{35}$

27) $2\frac{4}{35}$

28) $6\frac{5}{8}$

29) $2\frac{11}{21}$

30) $1\frac{3}{4}$

31) $1\frac{31}{32}$

32) $7\frac{11}{35}$

33) $2\frac{43}{72}$

34) $5\frac{7}{10}$

35) $8\frac{1}{2}$

36) $23\frac{1}{3}$

37) $3\frac{17}{26}$

38) $1\frac{39}{56}$

39) $2\frac{33}{64}$

40) $1\frac{49}{51}$

41) $1\frac{73}{75}$

42) $\frac{42}{55}$

Chapter 2:

Decimals

Math Topics that you'll learn in this Chapter:

- ✓ Comparing Decimals
- ✓ Rounding Decimals
- ✓ Adding and Subtracting Decimals
- ✓ Multiplying and Dividing Decimals

Comparing Decimals

- Decimal is a fraction written in a special form. For example, instead of writing $\frac{1}{2}$ you can write 0.5

- A Decimal Number contains a Decimal Point. It separates the whole number part from the fractional part of a decimal number.

- Let's review decimal place values: Example: 53.9861

    5: tens   3: ones   9: tenths

    8: hundredths   6: thousandths   1: tens thousandths

☑ To compare decimals, compare each digit of two decimals in the same place value. Start from left. Compare hundreds, tens, ones, tenth, hundredth, etc.

☑ To compare numbers, use these symbols:

Equal to $=$,   Less than $<$,   Greater than $>$
Greater than or equal $\geq$,   Less than or equal $\leq$

Examples:

1) Compare 0.60 and 0.06.

    Solution: 0.60 *is greater than* 0.06, because the tenth place of 0.60 is 6, but the tenth place of 0.06 is zero. Then: $0.60 > 0.06$

2) Compare 0.0815 and 0.815.

    Solution: 0.815 *is greater than* 0.0815, because the tenth place of 0.815 is 8, but the tenth place of 0.0815 is zero. Then: $0.0815 < 0.815$

                  ISEE Middle Level Math Prep 2020-2021

Rounding Decimals

☑ We can round decimals to a certain accuracy or number of decimal places. This is used to make calculation easier to do and results easier to understand, when exact values are not too important.

☑ First, you'll need to remember your place values: For example:

$$12.4869$$

1: tens	2: ones	4: tenths
8: hundredths	6: thousandths	9: tens thousandths

☑ To round a decimal, first find the place value you'll round to.

☑ Find the digit to the right of the place value you're rounding to. If it is 5 or bigger, add 1 to the place value you're rounding to and remove all digits on its right side. If the digit to the right of the place value is less than 5, keep the place value and remove all digits on the right.

Examples:

1) Round **1.9278** to the thousandth place value.

Solution: First look at the next place value to the right, (tens thousandths). It's 8 and it is greater than 5. Thus add 1 to the digit in the thousandth place. Thousandth place is 7. → $7 + 1 = 8$, then, the answer is 1.928

2) Round **9.4126** to the nearest hundredth.

Solution: First look at the digit to the right of hundredth (thousandths place value). It's 2 and it is less than 5, thus remove all the digits to the right of hundredth place. Then, the answer is 9.41

Adding and Subtracting Decimals

☑ Line up the decimal numbers.

☑ Add zeros to have same number of digits for both numbers if necessary.

☑ Remember your place values: For example:

$$73.5196$$

7: tens   3: ones   5: tenths

1: hundredths   9: thousandths   6: tens thousandths

☑ Add or subtract using column addition or subtraction.

Examples:

1) Add. $1.8 + 3.12$

Solution: First line up the numbers: $\begin{array}{r} 1.8 \\ +\,3.12 \\ \hline \end{array}$ → Add a zero to have same number of digits for both numbers. $\begin{array}{r} 1.80 \\ +\,3.12 \\ \hline \quad\quad \end{array}$ → Start with the hundredths place: $0 + 2 = 2$, $\begin{array}{r} 1.80 \\ +\,3.12 \\ \hline 2 \end{array}$ → Continue with tenths place: $8 + 1 = 9$, $\begin{array}{r} 1.80 \\ +\,3.12 \\ \hline .92 \end{array}$ → Add the ones place: $3 + 1 = 4$, $\begin{array}{r} 1.80 \\ +\,3.12 \\ \hline 4.92 \end{array}$

2) Find the difference. $3.67 - 2.23$

Solution: First line up the numbers: $\begin{array}{r} 3.67 \\ -\,2.23 \\ \hline \end{array}$ → Start with the hundredths place: $7 - 3 = 4$, $\begin{array}{r} 3.67 \\ -\,2.23 \\ \hline 4 \end{array}$ → Continue with tenths place. $6 - 2 = 4$, $\begin{array}{r} 3.67 \\ -\,2.23 \\ \hline .44 \end{array}$ → Subtract the ones place. $3 - 2 = 1$, $\begin{array}{r} 3.67 \\ -\,2.23 \\ \hline 1.44 \end{array}$

Multiplying and Dividing Decimals

For multiplying decimals:

☑ Ignore the decimal point and set up and multiply the numbers as you do with whole numbers.

☑ Count the total number of decimal places in both of the factors.

☑ Place the decimal point in the product.

For dividing decimals:

☑ If the divisor is not a whole number, move decimal point to right to make it a whole number. Do the same for dividend.

☑ Divide similar to whole numbers.

Examples:

1) Find the product. $0.81 \times 0.32 =$

Solution: Set up and multiply the numbers as you do with whole numbers. Line up the numbers: $\begin{array}{r} 81 \\ \times 32 \end{array}$ → Start with the ones place then continue with other digits → $\begin{array}{r} 81 \\ \times 32 \\ \hline 2{,}592 \end{array}$. Count the total number of decimal places in both of the factors. There are four decimals digits. (two for each factor 0.81 and 0.32) Then: $0.81 \times 0.32 = 0.2592$

2) Find the quotient. $1.60 \div 0.4 =$

Solution: The divisor is not a whole number. Multiply it by 10 to get 4: → $0.4 \times 10 = 4$

Do the same for the dividend to get 16. → $1.60 \times 10 = 1.6$

Now, divide: $16 \div 4 = 4$. The answer is 4.

Chapter 2: Practices

✍ *Compare. Use >, =, and <*

1) $0.88 \square 0.088$ 3) $0.99 \square 0.89$ 5) $1.58 \square 1.75$

2) $0.56 \square 0.57$ 4) $1.55 \square 1.65$ 6) $2.91 \square 2.85$

✍ *Round each decimal to the nearest whole number.*

7) 5.94   9) 9.7   11) 24.46

8) 16.47   10) 35.8   12) 12.5

✍ *Find the sum or difference.*

13) $43.15 + 23.65 =$ 15) $25.47 + 31.76 =$ 17) $45.53 + 18.95 =$

14) $56.74 - 22.43 =$ 16) $69.87 - 35.98 =$ 18) $25.13 - 18.72 =$

✍ *Find the product and quotient.*

19) $0.5 \times 0.8 =$ 21) $3.25 \times 2.2 =$ 23) $5.4 \times 0.6 =$

20) $6.4 \div 0.4 =$ 22) $8.4 \div 2.5 =$ 24) $1.42 \div 0.5 =$

Answers – Chapter 2

1) 0.88 > 0.088
2) 0.56 < 0.57
3) 0.99 > 0.89
4) 1.55 < 1.65
5) 1.58 < 1.75
6) 2.91 > 2.85
7) 6
8) 16
9) 10
10) 36
11) 24
12) 13

13) 66.8
14) 34.31
15) 57.23
16) 33.89
17) 64.48
18) 6.41
19) 0.4
20) 16
21) 7.15
22) 3.36
23) 3.24
24) 2.84

Chapter 3:

Integers and Order of

Operations

Math Topics that you'll learn in this Chapter:

- ✓ Adding and Subtracting Integers

- ✓ Multiplying and Dividing Integers

- ✓ Order of Operations

- ✓ Integers and Absolute Value

Adding and Subtracting Integers

☑ Integers include: zero, counting numbers, and the negative of the counting numbers. $\{\dots, -3, -2, -1, 0, 1, 2, 3, \dots\}$

☑ Add a positive integer by moving to the right on the number line. (you will get a bigger number)

☑ Add a negative integer by moving to the left on the number line. (you will get a smaller number)

☑ Subtract an integer by adding its opposite.

Examples:

1) Solve. $(-4) - (-5) =$

   Solution: Keep the first number and convert the sign of the second number to its opposite. (change subtraction into addition. Then: $(-4) + 5 = 1$

2) Solve. $11 + (8 - 19) =$

   Solution: First subtract the numbers in brackets, $8 - 19 = -11$.
   Then: $11 + (-11) = \rightarrow$ change addition into subtraction: $11 - 11 = 0$

3) Solve. $5 - (-14 - 3) =$

   Solution: First subtract the numbers in brackets, $-14 - 3 = -17$
   Then: $5 - (-17) = \rightarrow$ change subtraction into addition: $5 + 17 = 22$

4) Solve. $10 + (-6 - 15) =$

   Solution: First subtract the numbers in brackets, $-6 - 15 = -21$
   Then: $10 + (-21) = \rightarrow$ change addition into subtraction: $10 - 21 = -11$

Multiplying and Dividing Integers

Use following rules for multiplying and dividing integers:

☑ (negative) × (negative) = positive

☑ (negative) ÷ (negative) = positive

☑ (negative) × (positive) = negative

☑ (negative) ÷ (positive) = negative

☑ (positive) × (positive) = positive

☑ (positive) ÷ (negative) = negative

Examples:

1) Solve. $2 \times (-3) =$

Solution: Use this rule: (positive) × (negative) = negative.
Then: $(2) \times (-3) = -6$

2) Solve. $(-5) + (-27 \div 9) =$

Solution: First divided -27 by 9 , the numbers in brackets, use this rule:
(negative) ÷ (positive) = negative. Then: $-27 \div 9 = -3$
$(-5) + (-27 \div 9) = (-5) + (-3) = -5 - 3 = -8$

3) Solve. $(15 - 17) \times (-8) =$

Solution: First subtract the numbers in brackets, $15 - 17 = -2 \rightarrow (-2) \times (-8) =$

Now use this rule: (negative) × (negative) = positive
$(-2) \times (-8) = 16$

4) Solve. $(16 - 10) \div (-2) =$

Solution: First subtract the numbers in brackets, $16 - 10 = 6 \rightarrow (6) \div (-2) =$

Now use this rule: (positive) ÷ (negative) = negative
$(6) \div (-2) = -3$

Order of Operations

☑ In Mathematics, "operations" are addition, subtraction, multiplication, division, exponentiation (written as b^n), and grouping;

☑ When there is more than one math operation in an expression, use PEMDAS: (to memorize this rule, remember the phrase "Please Excuse My Dear Aunt Sally".)

- ❖ Parentheses
- ❖ Exponents
- ❖ Multiplication and Division (from left to right)
- ❖ Addition and Subtraction (from left to right)

Examples:

1) Calculate. $(3 + 5) \div (3^2 \div 9) =$

   Solution: First simplify inside parentheses: $(8) \div (9 \div 9) = (8) \div (1)$, Then: $(8) \div (1) = 8$

2) Solve. $(7 \times 8) - (12 - 4) =$

   Solution: First calculate within parentheses: $(7 \times 8) - (12 - 4) = (56) - (8)$, Then: $(56) - (8) = 48$

3) Calculate. $-2[(8 \times 9) \div (2^2 \times 2)] =$

   Solution: First calculate within parentheses: $-2[(72) \div (4 \times 2)] = -2[(72) \div (8)] = -2[9]$ multiply -2 and 9. Then: $-2[9] = -18$

4) Solve. $(14 \div 7) + (-13 + 8) =$

   Solution: First calculate within parentheses: $(14 \div 7) + (-13 + 8) = (2) + (-5)$

   Then: $(2) - (5) = -3$

Integers and Absolute Value

☑ The absolute value of a number is its distance from zero, in either direction, on the number line. For example, the distance of 9 and -9 from zero on number line is 9.

☑ The absolute value of an integer is the numerical value without its sign. (negative or positive)

☑ The vertical bar is used for absolute value as in $|x|$.

☑ The absolute value of a number is never negative; because it only shows, "how far the number is from zero".

Examples:

1) Calculate. $|12 - 4| \times 4 =$

Solution: First solve $|12 - 4|$, $\rightarrow |12 - 4| = |8|$, the absolute value of 8 is 8, $|8| = 8$
Then: $8 \times 4 = 32$

2) Solve. $\frac{|-16|}{4} \times |3 - 8| =$

Solution: First find $|-16|$, $\rightarrow$ the absolute value of -16 is 16, then: $|-16| = 16$,
$\frac{16}{4} \times |3 - 8| =$
Now, calculate $|3 - 8|$, $\rightarrow |3 - 8| = |-5|$, the absolute value of -5 is 5. $|-5| = 5$
Then: $\frac{16}{4} \times 5 = 4 \times 5 = 20$

3) Solve. $|9 - 3| \times \frac{|-3 \times 8|}{6} =$

Solution: First calculate $|9 - 3|$, $\rightarrow |9 - 3| = |6|$, the absolute value of 6 is 6, $|6| = 6$. Then:
$6 \times \frac{|-3 \times 8|}{6}$
Now calculate $|-3 \times 8|$, $\rightarrow |-3 \times 8| = |-24|$, the absolute value of -24 is 24, $|-24| = 24$
Then: $6 \times \frac{24}{6} = 6 \times 4 = 24$

Chapter 3: Practices

✍ *Find each sum or difference.*

1) $18 + (-5) =$

2) $(-16) + 24 =$

3) $(-12) + (-9) =$

4) $14 + (-8) + 6 =$

5) $24 + (-10 - 7) =$

6) $(-15) + (-6 + 12) =$

✍ *Find each product or quotient.*

7) $8 \times (-6) =$

8) $(-12) \div (-3) =$

9) $(-4) \times (-7) \times 2 =$

10) $3 \times (-5) \times (-6) =$

11) $(-7 - 37) \div (-11) =$

12) $(8 - 6) \times (-24) =$

✍ *Evaluate each expression.*

13) $8 + (3 \times 7) =$

14) $(18 \times 2) - 14 =$

15) $(15 - 7) + (2 \times 6) =$

16) $(8 + 4) \div (2^3 \div 2) =$

17) $2[(6 \times 3) \div (3^2 \times 2)] =$

18) $-3[(8 \times 2^2) \div (8 \times 2)] =$

✍ *Find the answers.*

19) $|-6| + |9 - 12| =$

20) $|8| - |7 - 19| + 1 =$

21) $\frac{|-40|}{8} \times \frac{|-15|}{5} =$

22) $|7 \times -5| \times \frac{|-32|}{8} =$

23) $\frac{|-121|}{11} - |-8 \times 2| =$

24) $\frac{|-3 \times -6|}{9} \times \frac{|4 \times -6|}{8} =$

Answers – Chapter 3

1) 13
2) 8
3) −21
4) 12
5) 7
6) −9
7) −48
8) 4
9) 56
10) 90
11) 4
12) −48

13) 29
14) 22
15) 20
16) 3
17) 2
18) −6
19) 9
20) −3
21) 15
22) 140
23) −5
24) 6

Chapter 4:
Ratios and Proportions

Math Topics that you'll learn in this Chapter:

- ✓ Simplifying Ratios
- ✓ Proportional Ratios
- ✓ Similarity and Ratios

Simplifying Ratios

☑ Ratios are used to make comparisons between two numbers.

☑ Ratios can be written as a fraction, using the word "to", or with a colon.
Example: $\frac{3}{4}$ or "3 to 4" or $3:4$

☑ You can calculate equivalent ratios by multiplying or dividing both sides
of the ratio by the same number.

Examples:

1) Simplify. $9:3 =$

   Solution: Both numbers 9 and 3 are divisible by 3 , $\Rightarrow 9 \div 3 = 3,$ $\quad 3 \div 3 = 1$, Then:
   $9:3 = 3:1$

2) Simplify. $\frac{24}{44} =$

   Solution: Both numbers 24 and 44 are divisible by 4, $\Rightarrow 24 \div 4 = 6, 44 \div 4 = 11$, Then:
   $\frac{24}{44} = \frac{6}{11}$

3) There are 36 students in a class and 16 of them are girls. Write the ratio of
   girls to boys.

   Solution: Subtract 16 from 36 to find the number of boys in the class. $36 - 16 = 20$. There
   are 20 boys in the class. So, ratio of girls to boys is $16:20$. Now, simplify this ratio. Both 20
   and 16 are divisible by 4. Then: $20 \div 4 = 5$, and $16 \div 4 = 4$. In simplest form, this ratio is
   $4:5$

4) A recipe calls for butter and sugar in the ratio $3:4$. If you're using 9 cups of
   butter, how many cups of sugar should you use?

   Solution: Since, you use 9 cups of butter, or 3 times as much, you need to multiply the
   amount of sugar by 3. Then: $4 \times 3 = 12$. So, you need to use 12 cups of sugar. You can
   solve this using equivalent fractions: $\frac{3}{4} = \frac{9}{12}$

Proportional Ratios

☑ Two ratios are proportional if they represent the same relationship.

☑ A proportion means that two ratios are equal. It can be written in two

ways: $\dfrac{a}{b} = \dfrac{c}{d}$ $a : b = c : d$

☑ The proportion $\dfrac{a}{b} = \dfrac{c}{d}$ can be written as: $a \times d = c \times b$

Examples:

1) Solve this proportion for x. $\dfrac{3}{7} = \dfrac{12}{x}$

Solution: Use cross multiplication: $\dfrac{3}{7} = \dfrac{12}{x} \Rightarrow 3 \times x = 7 \times 12 \Rightarrow 3x = 84$

Divide both sides by 3 to find x: $x = \dfrac{84}{3} \Rightarrow x = 28$

2) If a box contains red and blue balls in ratio of $3:7$ red to blue, how many red balls are there if 49 blue balls are in the box?

Solution: Write a proportion and solve. $\dfrac{3}{7} = \dfrac{x}{49}$

Use cross multiplication: $3 \times 49 = 7 \times x \Rightarrow 147 = 7x$

Divide to find x: $x = \dfrac{147}{7} \Rightarrow x = 21$. There are 21 red balls in the box.

3) Solve this proportion for x. $\dfrac{2}{9} = \dfrac{12}{x}$

Solution: Use cross multiplication: $\dfrac{2}{9} = \dfrac{12}{x} \Rightarrow 2 \times x = 9 \times 12 \Rightarrow 2x = 108$

Divide to find x: $x = \dfrac{108}{2} \Rightarrow x = 54$

4) Solve this proportion for x. $\dfrac{6}{7} = \dfrac{18}{x}$

Solution: Use cross multiplication: $\dfrac{6}{7} = \dfrac{18}{x} \Rightarrow 6 \times x = 7 \times 18 \Rightarrow 6x = 126$

Divide to find x: $x = \dfrac{126}{6} \Rightarrow x = 21$

Similarity and Ratios

☑ Two figures are similar if they have the same shape.

☑ Two or more figures are similar if the corresponding angles are equal, and the corresponding sides are in proportion.

Examples:

1) Following triangles are similar. What is the value of unknown side?

Solution: Find the corresponding sides and write a proportion.

$\frac{5}{10} = \frac{4}{x}$. Now, use cross product to solve for x:

$\frac{5}{10} = \frac{4}{x} \rightarrow 5 \times x = 10 \times 4 \rightarrow 5x = 40$. Divide

both sides by 5. Then: $5x = 40 \rightarrow \frac{5x}{5} = \frac{40}{5} \rightarrow x = 8$

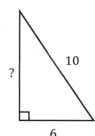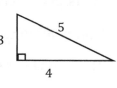

The missing side is 8.

2) Two rectangles are similar. The first is 6 feet wide and 20 feet long. The second is 15 feet wide. What is the length of the second rectangle?

Solution: Let's put x for the length of the second rectangle. Since two rectangles are similar, their corresponding sides are in proportion. Write a proportion and solve for the missing number. $\frac{6}{15} = \frac{20}{x} \rightarrow 6x = 15 \times 20 \rightarrow 6x = 300 \rightarrow x = \frac{300}{6} = 50$

The length of the second rectangle is 50 feet.

Chapter 4: Practices

✍ *Reduce each ratio.*

1) $9 : 18 = $ ___ $:$ ___

2) $6 : 54 = $ ___ $:$ ___

3) $28 : 49 = $ ___ $:$ ___

4) Bob has 12 red cards and 20 green cards. What is the ratio of Bob's red cards to his green cards? _____

5) In a party, 10 soft drinks are required for every 12 guests. If there are 252 guests, how many soft drinks is required? _____

6) In Jack's class, 18 of the students are tall and 10 are short. In Michael's class 54 students are tall and 30 students are short. Which class has a higher ratio of tall to short students? _____

✍ *Solve each proportion.*

7) $\dfrac{3}{7} = \dfrac{18}{x} , x = $ ____

8) $\dfrac{5}{9} = \dfrac{x}{108} , x = $ ____

9) $\dfrac{2}{13} = \dfrac{8}{x} , x = $ ____

10) $\dfrac{4}{10} = \dfrac{6}{x} , x = $ ____

11) $\dfrac{8}{20} = \dfrac{x}{65} , x = $ ____

12) $\dfrac{6}{15} = \dfrac{14}{x} , x = $ ____

✍ *Solve each problem.*

13) Two rectangles are similar. The first is $8\,feet$ wide and $22\,feet$ long. The second is $12\,feet$ wide. What is the length of the second rectangle? _____

14) Two rectangles are similar. One is $3.2\,meters$ by $8\,meters$. The longer side of the second rectangle is $34.5\,meters$. What is the other side of the second rectangle? _____

Answers – Chapter 4

1) 1 : 2
2) 1 : 9
3) 4 : 7
4) 3 : 5
5) 210
6) The ratio for both classes is 9 to 5.
7) 42

8) 60
9) 52
10) 15
11) 26
12) 35
13) 33 feet
14) 13.8 meters

Chapter 5:

Percentage

Math Topics that you'll learn in this Chapter:

- ✓ Percentage Calculations
- ✓ Percent Problems
- ✓ Percent of Increase and Decrease
- ✓ Discount, Tax and Tip
- ✓ Simple Interest

Percent Problems

☑ Percent is a ratio of a number and 100. It always has the same denominator, 100. Percent symbol is "%".

☑ Percent means "per 100". So, 20% is 20/100.

☑ In each percent problem, we are looking for the base, or part or the percent.

☑ Use the following equations to find each missing section in a percent problem:

- Base = Part ÷ Percent
- Part = Percent × Base
- Percent = Part ÷ Base

Examples:

1) What is 25% of 60?

Solution: In this problem, we have percent (25%) and base (60) and we are looking for the "part".

Use this formula: $part = percent \times base$. Then: $part = 25\% \times 60 = \frac{25}{100} \times 60 = 0.25 \times 60 = 15$.

The answer:

25% of 60 is 15.

2) 20 is what percent of 400?

Solution: In this problem, we are looking for the percent. Use this equation:

$Percent = Part \div Base \rightarrow Percent = 20 \div 400 = 0.05 = 5\%$.

Then: 20 is 5 percent of 400.

Percent of Increase and Decrease

☑ Percent of change (increase or decrease) is a mathematical concept that represents the degree of change over time.

☑ To find the percentage of increase or decrease:

1. New Number – Original Number

2. The result ÷ Original Number × 100

☑ Or use this formula: Percent of change = $\frac{new\ number - original\ number}{original\ number} \times 100$

☑ Note: If your answer is a negative number, then this is a percentage decrease. If it is positive, then this is a percentage increase.

Examples:

1) The price of a shirt increases from $20 to $30. What is the percentage increase?
   Solution: First find the difference: $30 - 20 = 10$
   Then: $10 \div 20 \times 100 = \frac{10}{20} \times 100 = 50$. The percentage increase is 50. It means that the price of the shirt increased 50%.

2) The price of a table increased from $25 to $40. What is the percent of increase?
   Solution: Use percentage formula: $Percent\ of\ change = \frac{new\ number - original\ number}{original\ number} \times$
   $100 = \frac{40-25}{25} \times 100 = \frac{15}{25} \times 100 = 0.6 \times 100 = 60$. The percentage increase is 60. It means that the price of the table increased 60%.

3) The population of a town was 50,000 in the 2000 census and 40,000 in the 2010 census. By what percent did the population decrease?
   Solution: Use percentage formula:
   $Percent\ of\ change = \frac{new\ number - original\ number}{original\ number} \times 100 = \frac{40,000-50,000}{50,000} \times 100 = \frac{-10,000}{50,000} \times$
   $100 = -0.2 \times 100 = -20$. The population of the town decreased by 20%.

Discount, Tax and Tip

☑ To find discount: Multiply the regular price by the rate of discount

☑ To find selling price: Original price – discount

☑ To find tax: Multiply the tax rate to the taxable amount (income, property value, etc.)

☑ To find tip, multiply the rate to the selling price.

Examples:

1) With an 10% discount, Ella was able to save $45 on a dress. What was the original price of the dress?

   Solution: let x be the original price of the dress. Then: $10\% \ of \ x = 45$. Write an equation and solve for x: $0.10 \times x = 45 \rightarrow x = \frac{45}{0.10} = 450$. The original price of the dress was $450.

2) Sophia purchased a new computer for a price of $950 at the Apple Store. What is the total amount her credit card is charged if the sales tax is 7%?

   Solution: The taxable amount is $950, and the tax rate is 7%. Then: $Tax = 0.07 \times 950 = 66.50$

   $Final \ price = Selling \ price + Tax \rightarrow final \ price = \$950 + \$66.50 = \$1,016.50$

3) Nicole and her friends went out to eat at a restaurant. If their bill was $80.00 and they gave their server a 15% tip, how much did they pay altogether?

   Solution: First find the tip. To find tip, multiply the rate to the bill amount.

   $Tip = 80 \times 0.15 = 12$. The final price is: $\$80 + \$12 = \$92$

Simple Interest

☑ Simple Interest: The charge for borrowing money or the return for lending it.

☑ Simple interest is calculated on the initial amount (principal).

☑ To solve a simple interest problem, use this formula:

Interest = principal x rate x time   ($I = p \times r \times t = prt$)

Examples:

1) Find simple interest for $300 investment at 6% for 5 years.

   Solution: Use Interest formula: $I = prt$ ($P = \$300$, r = 6% $= \frac{6}{100} = 0.06$ and $t = 5$)
   Then: $I = 300 \times 0.06 \times 5 = \$90$

2) Find simple interest for $1,600 at 5% for 2 years.

   Solution: Use Interest formula: $I = prt$ ($P = \$1,600$, r = 5% $= \frac{5}{100} = 0.05$ and $t = 2$)
   Then: $I = 1,600 \times 0.05 \times 2 = \$160$

3) Andy received a student loan to pay for his educational expenses this year. What is the interest on the loan if he borrowed $6,500 at 8% for 6 years?
   Solution: Use Interest formula: $I = prt$. $P = \$6,500$, r = 8% $= 0.08$ and $t = 6$

   Then: $I = 6,500 \times 0.08 \times 8 = \$3,120$

4) Bob is starting his own small business. He borrowed $10,000 from the bank at a 6% rate for 6 months. Find the interest Bob will pay on this loan.
   Solution: Use Interest formula: $I = prt$. $P = \$10,000$, r = 6% $= 0.06$ and $t = 0.5$ (6 months is half year). Then: $I = 10,000 \times 0.06 \times 0.5 = \$300$

Chapter 5: Practices

✍ *Solve each problem.*

1) 15 is what percent of 60? ____%

2) 18 is what percent of 24? ____%

3) 25 is what percent of 500? ____%

4) 14 is what percent of 280? ____%

5) 45 is what percent of 180? ____%

6) 70 is what percent of 350? ____%

✍ *Solve each percent of change word problem.*

7) Bob got a raise, and his hourly wage increased from $15 to $18. What is the percent increase? _____ %

8) The price of a pair of shoes increases from $40 to $66. What is the percent increase? ____ %

9) At a coffeeshop, the price of a cup of coffee increased from $1.20 to $1.44. What is the percent increase in the cost of the coffee? _____ %

✍ *Find the selling price of each item.*

10) Original price of a computer: $650

Tax: 8%, Selling price: $_____

11) Original price of a laptop: $480

Tax: 15%, Selling price: $_____

12) Nicolas hired a moving company. The company charged $400 for its services, and Nicolas gives the movers a 15% tip. How much does Nicolas tip the movers? $_____

13) Mason has lunch at a restaurant and the cost of his meal is $30. Mason wants to leave a 20% tip. What is Mason's total bill including tip? $_____

✎ *Determine the simple interest for these loans.*

14) $440 *at* 5% *for* 6 *years*. $__

15) $460 *at* 2.5% *for* 4 *years*. $_

16) $500 *at* 3% *for* 5 *years*. $__

17) $550 at 9% for 2 years. $__

18) A new car, valued at $28,000, depreciates at 9% per year. What is the value of the car one year after purchase? $_____

19) Sara puts $4,000 into an investment yielding 5% annual simple interest; she left the money in for five years. How much interest does Sara get at the end of those five years? $_____

Answers – Chapter 5

1) 25%
2) 75%
3) 5%
4) 5%
5) 25%
6) 20%
7) 20%
8) 65%
9) 20%
10) $702.00

11) $552.00
12) $60.00
13) $36.00
14) $132
15) $46
16) $75
17) $99
18) $25,480.00
19) $1,000.00

Chapter 6:

Expressions and Variables

Math Topics that you'll learn in this Chapter:

- ✓ Simplifying Variable Expressions
- ✓ Simplifying Polynomial Expressions
- ✓ The Distributive Property
- ✓ Evaluating One Variable
- ✓ Evaluating Two Variables

Simplifying Variable Expressions

☑ In algebra, a variable is a letter used to stand for a number. The most common letters are: $x, y, z, a, b, c, m, and\ n$.

☑ Algebraic expression is an expression contains integers, variables, and the math operations such as addition, subtraction, multiplication, division, etc.

☑ In an expression, we can combine "like" terms. (values with same variable and same power)

Examples:

1) Simplify. $(2x + 3x + 4) =$

   Solution: In this expression, there are three terms: $2x, 3x$, and 4. Two terms are "like terms": $2x$ and $3x$. Combine like terms. $2x + 3x = 5x$. Then: $(2x + 3x + 4) = 5x + 4$ (remember you cannot combine variables and numbers.)

2) Simplify. $12 - 3x^2 + 5x + 4x^2 =$

   Solution: Combine "like" terms: $-3x^2 + 4x^2 = x^2$. Then:
   $12 - 3x^2 + 5x + 4x^2 = 12 + x^2 + 5x$. Write in standard form (biggest powers first):
   $12 + x^2 + 5x = x^2 + 5x + 12$

3) Simplify. $(10x^2 + 2x^2 + 3x) =$

   Solution: Combine like terms. Then: $(10x^2 + 2x^2 + 3x) = 12x^2 + 3x$

4) Simplify. $15x - 3x^2 + 9x + 5x^2 =$

   Solution: Combine "like" terms: $15x + 9x = 24x$, and $-3x^2 + 5x^2 = 2x^2$

   Then: $15x - 3x^2 + 9x + 5x^2 = 24x + 2x^2$. Write in standard form (biggest powers first): $24x + 2x^2 = 2x^2 + 24x$

Simplifying Polynomial Expressions

☑ In mathematics, a polynomial is an expression consisting of variables and coefficients that involves only the operations of addition, subtraction, multiplication, and non-negative integer exponents of variables. $P(x) = a_n x^n + a_{n-1} x^{n-1} + \ldots + a_2 x^2 + a_1 x + a_0$

☑ Polynomials must always be simplified as much as possible. It means you must add together any like terms. (values with same variable and same power)

Examples:

1) Simplify this Polynomial Expressions. $x^2 - 5x^3 + 2x^4 - 4x^3$

   Solution: Combine "like" terms: $-5x^3 - 4x^3 = -9x^3$

   Then: $x^2 - 5x^3 + 2x^4 - 4x^3 = x^2 - 9x^3 + 2x^4$

   Now, write the expression in standard form: $2x^4 - 9x^3 + x^2$

2) Simplify this expression. $(2x^2 - x^3) - (x^3 - 4x^2) =$

   Solution: First use distributive property: $\rightarrow$ multiply $(-)$ into $(x^3 - 4x^2)$

   $(2x^2 - x^3) - (x^3 - 4x^2) = 2x^2 - x^3 - x^3 + 4x^2$

   Then combine "like" terms: $2x^2 - x^3 - x^3 + 4x^2 = 6x^2 - 2x^3$

   And write in standard form: $6x^2 - 2x^3 = -2x^3 + 6x^2$

3) Simplify. $4x^4 - 5x^3 + 15x^4 - 12x^3 =$

   Solution: Combine "like" terms: $-5x^3 - 12x^3 = -17x^3$ and $4x^4 + 15x^4 = 19x^4$

   Then: $4x^4 - 5x^3 + 15x^4 - 12x^3 = 19x^4 - 17x^3$

The Distributive Property

- ☑ The distributive property (or the distributive property of multiplication over addition and subtraction) simplifies and solves expressions in the form of: $a(b + c)$ or $a(b - c)$

- ☑ The distributive property is multiplying a term outside the parentheses by the terms inside.

- ☑ Distributive Property rule: $a(b + c) = ab + ac$

Examples:

1) *Simply using distributive property.* $(-4)(x - 5)$

   Solution: Use Distributive Property rule: $a(b + c) = ab + ac$
   $(-4)(x - 5) = (-4 \times x) + (-4) \times (-5) = -4x + 20$

2) *Simply.* $(3)(2x - 4)$

   Solution: Use Distributive Property rule: $a(b + c) = ab + ac$
   $(3)(2x - 4) = (3 \times 2x) + (3) \times (-4) = 6x - 12$

3) *Simply.* $(-3)(3x - 5) + 4x$

   Solution: First, simplify $(-3)(3x - 5)$ using distributive property.
   Then: $(-3)(3x - 5) = -9x + 15$
   Now combine like terms: $(-3)(3x - 5) + 4x = -9x + 15 + 4x$
   In this expression, $-9x$ and $4x$ are "like terms" and we can combine them.
   $-9x + 4x = -5x$. Then: $-9x + 15 + 4x = -5x + 15$

Evaluating One Variable

☑ To evaluate one variable expressions, find the variable and substitute a number for that variable.

☑ Perform the arithmetic operations.

Examples:

1) *Calculate this expression for* x = 3.　　$15 - 3x$

   Solution: First substitute 3 for x

   Then: $15 - 3x = 15 - 3(3)$

   Now, use order of operation to find the answer: $15 - 3(3) = 15 - 9 = 6$

2) *Evaluate this expression for* x = 1.　　$5x - 12$

   Solution: First substitute 1 for x, then:

   $5x - 12 = 5(1) - 12$

   Now, use order of operation to find the answer: $5(1) - 12 = 5 - 12 = -7$

3) *Find the value of this expression when* x = 5.　$25 - 4x$

   Solution: First substitute 5 for x, then:

   $25 - 4x = 25 - 4(5) = 25 - 20 = 5$

4) *Solve this expression for* x = -2.　　$12 + 3x$

   Solution: Substitute -2 for x, then: $12 + 3x = 12 + 3(-2) = 12 - 6 = 6$

Evaluating Two Variables

☑ To evaluate an algebraic expression, substitute a number for each variable.

☑ Perform the arithmetic operations to find the value of the expression.

Examples:

1) *Calculate this expression for* $a = 3$ *and* $b = -2$. $3a - 6b$

   Solution: First substitute 3 for a, and -2 for b , then:
   $$3a - 6b = 3(3) - 6(-2)$$
   Now, use order of operation to find the answer: $3(3) - 6(-2) = 9 + 12 = 21$

2) *Evaluate this expression for* $x = 3$ *and* $y = 1$. $3x + 5y$

   Solution: Substitute 3 for x, and 1 for y , then:
   $$3x + 5y = 3(3) + 5(1) = 9 + 5 = 14$$

3) *Find the value of this expression when* $a = 1$ *and* $b = 2$. $5(3a - 2b)$

   Solution: Substitute 1 for a, and 2 for b , then:
   $$5(3a - 2b) = 15a - 10b = 15(1) - 10(2) = 15 - 20 = -5$$

4) *Solve this expression.* $4x - 3y$, $x = 3$, $y = 5$

   Solution: Substitute 3 for x, and 5 for y and simplify. Then: $4x - 3y = 4(3) - 3(5) = 12 - 15 = -3$

Chapter 6: Practices

✎ **Simplify each expression.**

1) $(6x - 4x + 8 + 6) =$

2) $(-14x + 26x - 12) =$

3) $(24x - 6 - 18x + 3) =$

4) $5 + 8x^2 - 9 =$

5) $7x - 4x^2 + 6x =$

6) $15x^2 - 3x - 6x^2 + 4 =$

✎ **Simplify each polynomial.**

7) $2x^2 + 5x^3 - 7x^2 + 12x =$ _____

8) $2x^4 - 5x^5 + 8x^4 - 8x^2 =$ _____

9) $5x^3 + 15x - x^2 - 2x^3 =$ _____

10) $(8x^3 - 6x^2) + (9x^2 - 10x) =$ _____

11) $(12x^4 + 4x^3) - (8x^3 - 2x^4) =$ _____

12) $(9x^5 - 7x^3) - (5x^3 + x^2) =$ _____

✎ **Use the distributive property to simply each expression.**

13) $4(5 + 6x) =$

14) $5(8 - 4x) =$

15) $(-6)(2 - 9x) =$

16) $(-7)(6x - 4) =$

17) $(3x + 12)4 =$

18) $(8x - 5)(-3) =$

✎ **Evaluate each expression using the value given.**

19) $8 - x, x = -3$

20) $x + 12, x = -6$

21) $5x - 3, x = 2$

22) $4 - 6x, x = 1$

23) $3x + 1, x = -2$

24) $15 - 2x, x = 5$

✎ Evaluate each expression using the values given.

25) $4x - 2y, \ x = 4, y = -2$

26) $6a + 3b, \ a = 2, b = 4$

27) $12x - 5y - 8, \ x = 2, y = 3$

28) $-7a + 3b + 9, \ a = 4, b = 6$

29) $2x + 14 + 4y, \ x = 6, y = 8$

30) $4a - (5a - b) + 5, a = 4, b = 6$

Answers – Chapter 6

1) $2x + 14$
2) $12x - 12$
3) $6x - 3$
4) $8x^2 - 4$
5) $-4x^2 + 13x$
6) $9x^2 - 3x + 4$

7) $5x^3 - 5x^2 + 12x$
8) $-5x^5 + 10x^4 - 8x^2$
9) $3x^3 - x^2 + 15x$

10) $8x^3 + 3x^2 - 10x$
11) $14x^4 - 4x^3$
12) $9x^5 - 12x^3 - x^2$

13) $24x + 20$
14) $-20x + 40$
15) $54x - 12$
16) $-42x + 28$
17) $12x + 48$
18) $-24x + 15$

19) 11
20) 6
21) 7
22) -2
23) -5
24) 5

25) 20
26) 24
27) 1
28) -1
29) 58
30) 7

Chapter 7:

Equations and Inequalities

Math Topics that you'll learn in this Chapter:

- ✓ One–Step Equations

- ✓ Multi–Step Equations

- ✓ System of Equations

- ✓ Graphing Single–Variable Inequalities

- ✓ One–Step Inequalities

- ✓ Multi–Step Inequalities

One–Step Equations

☑ The values of two expressions on both sides of an equation are equal. Example: $ax = b$. In this equation, ax is equal to b.

☑ Solving an equation means finding the value of the variable.

☑ You only need to perform one Math operation in order to solve the one-step equations.

☑ To solve one-step equation, find the inverse (opposite) operation is being performed.

☑ The inverse operations are:

- Addition and subtraction
- Multiplication and division

Examples:

1) *Solve this equation for x.* $3x = 18, x = ?$

   Solution: Here, the operation is multiplication (variable x is multiplied by 3) and its inverse operation is division. To solve this equation, divide both sides of equation by 3:

   $$3x = 18 \rightarrow \frac{3x}{3} = \frac{18}{3} \rightarrow x = 6$$

2) *Solve this equation.* $x + 15 = 0$, $x = ?$

   Solution: In this equation 15 is added to the variable x. The inverse operation of addition is subtraction. To solve this equation, subtract 15 from both sides of the equation: $x + 15 - 15 = 0 - 15$. Then simplify: $x + 15 - 15 = 0 - 15 \rightarrow x = -15$

3) *Solve this equation for x.* $x - 23 = 0$

   Solution: Here, the operation is subtraction and its inverse operation is addition. To solve this equation, add 23 to both sides of the equation: $x + 23 - 23 = 0 - 23 \rightarrow x = -23$

Multi–Step Equations

☑ To solve a multi-step equation, combine "like" terms on one side.

☑ Bring variables to one side by adding or subtracting.

☑ Simplify using the inverse of addition or subtraction.

☑ Simplify further by using the inverse of multiplication or division.

☑ Check your solution by plugging the value of the variable into the original equation.

Examples:

1) *Solve this equation for x.* $3x + 6 = 16 - 2x$

Solution: First bring variables to one side by adding $2x$ to both sides. Then:

$3x + 6 = 16 - 2x \rightarrow 3x + 6 + 2x = 16 - 2x + 2x$. Simplify: $5x + 6 = 16$

Now, subtract 6 from both sides of the equation: $5x + 6 - 6 = 16 - 6 \rightarrow 5x = 10 \rightarrow$

Divide both sides by 5: $5x = 10 \rightarrow \frac{5x}{5} = \frac{10}{5} \rightarrow x = 2$

Let's check this solution by substituting the value of 2 for x in the original equation:

$x = 2 \rightarrow 3x + 6 = 16 - 2x \rightarrow 3(2) + 6 = 16 - 2(2) \rightarrow 6 + 6 = 16 - 4 \rightarrow 12 = 12$

The answer $x = 2$ is correct.

2) *Solve this equation for x.* $-4x + 4 = 16$

Solution: Subtract 4 from both sides of the equation. $-4x + 4 - 4 = 16 - 4 \rightarrow -4x = 12$

Divide both sides by -4, then: $-4x = 12 \rightarrow \frac{-4x}{-4} = \frac{12}{-4} \rightarrow x = -3$

Now, check the solution: $x = -3 \rightarrow -4x + 4 = 16 \rightarrow -4(-3) + 4 = 16 \rightarrow 16 = 16$

The answer $x = -2$ is correct.

System of Equations

☑ A system of equations contains two equations and two variables. For example, consider the system of equations: $x - y = 1, x + y = 5$

☑ The easiest way to solve a system of equations is using the elimination method. The elimination method uses the addition property of equality. You can add the same value to each side of an equation.

☑ For the first equation above, you can add $x + y$ to the left side and 5 to the right side of the first equation: $x - y + (x + y) = 1 + 5$. Now, if you simplify, you get: $x - y + (x + y) = 1 + 5 \rightarrow 2x = 6 \rightarrow x = 3$. Now, substitute 3 for the x in the first equation: $3 - y = 1$. By solving this equation, $y = 2$

Example:

What is the value of x + y *in this system of equations?* $\begin{cases} x + 2y = 6 \\ 2x - y = -8 \end{cases}$

Solution: Solving a System of Equations by Elimination:

Multiply the first equation by (-2), then add it to the second equation.

$$\begin{matrix} -2(x + 2y = 6) \\ 2x - y = -8 \end{matrix} \Rightarrow \begin{matrix} -2x - 4y = -12 \\ 2x - y = -8 \end{matrix} \Rightarrow -5y = -20 \Rightarrow y = 4$$

Plug in the value of y into one of the equations and solve for x.

$x + 2(4) = 6 \Rightarrow x + 8 = 6 \Rightarrow x = 6 - 8 \Rightarrow x = -2$

Thus, $x + y = -2 + 4 = 2$

Graphing Single–Variable Inequalities

☑ An inequality compares two expressions using an inequality sign.

☑ Inequality signs are: "less than" <, "greater than" >, "less than or equal to" ≤, and "greater than or equal to" ≥.

☑ To graph a single-variable inequality, find the value of the inequality on the number line.

☑ For less than (<) or greater than (>) draw open circle on the value of the variable. If there is an equal sign too, then use filled circle.

☑ Draw an arrow to the right for greater or to the left for less than.

Examples:

1) *Draw a graph for this inequality.* $x > 3$

Solution: Since, the variable is greater than 3, then we need to find 3 in the number line and draw an open circle on it.

Then, draw an arrow to the right.

2) *Graph this inequality.* $x \leq -4$.

Solution: Since, the variable is less than or equal to -4, then we need to find -4 in the number line and draw a filled circle on it. Then, draw an arrow to the left.

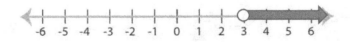

ISEE Middle Level Math Prep 2020-2021

One–Step Inequalities

✓ An inequality compares two expressions using an inequality sign.

✓ Inequality signs are: "less than" $<$, "greater than" $>$, "less than or equal to" $\leq$, and "greater than or equal to" $\geq$.

✓ You only need to perform one Math operation in order to solve the one-step inequalities.

✓ To solve one-step inequalities, find the inverse (opposite) operation is being performed.

✓ For dividing or multiplying both sides by negative numbers, flip the direction of the inequality sign.

Examples:

1) *Solve this inequality for* x. $x + 3 \geq 4$

   Solution: The inverse (opposite) operation of addition is subtraction. In this inequality, 3 is added to x. *To isolate* x *we need to* subtract 3 from both sides of the inequality. Then:

   $x + 3 \geq 4 \rightarrow x + 3 - 3 \geq 4 - 3 \rightarrow x \geq 1$. The solution is: $x \geq 1$

2) *Solve the inequality.* $x - 5 > -4$.

   Solution: 5 is subtracted from x. Add 5 to both sides. $x - 5 > -4 \rightarrow x - 5 + 5 > -4 + 5 \rightarrow x > 1$

3) *Solve.* $2x \leq -4$.

   Solution: 2 is multiplied to x. Divide both sides by 2. Then: $2x \leq -4 \rightarrow \frac{2x}{2} \leq \frac{-4}{2} \rightarrow x \leq -2$

4) *Solve.* $-6x \leq 12$.

   Solution: -6 is multiplied to x. Divide both sides by -6. Remember when dividing or multiplying both sides of an inequality by negative numbers, flip the direction of the inequality sign. Then:

   $$-6x \leq 12 \rightarrow \frac{-6x}{-6} \geq \frac{12}{-6} \rightarrow x \geq -2$$

Multi–Step Inequalities

☑ To solve a multi-step inequality, combine "like" terms on one side.

☑ Bring variables to one side by adding or subtracting.

☑ Isolate the variable.

☑ Simplify using the inverse of addition or subtraction.

☑ Simplify further by using the inverse of multiplication or division.

☑ For dividing or multiplying both sides by negative numbers, flip the direction of the inequality sign.

Examples:

1) *Solve this inequality.* $2x - 3 \leq 5$

   Solution: In this inequality, 3 is subtracted from $2x$. The inverse of subtraction is addition. Add 3 to both sides of the inequality: $2x - 3 + 3 \leq 5 + 3 \rightarrow 2x \leq 8$

   Now, divide both sides by 2. Then: $2x \leq 8 \rightarrow \frac{2x}{2} \leq \frac{8}{2} \rightarrow x \leq 4$

   The solution of this inequality is $x \leq 4$.

2) *Solve this inequality.* $3x + 9 < 12$

   Solution: First subtract 9 from both sides: $3x + 9 - 9 < 12 - 9$

   Then simplify: $3x + 9 - 9 < 12 - 9 \rightarrow 3x < 3$

   Now divide both sides by 3: $\frac{3x}{3} < \frac{3}{3} \rightarrow x < 1$

3) *Solve this inequality.* $-2x + 4 \geq 6$

   First subtract 4 from both sides: $-2x + 4 - 4 \geq 6 - 4 \rightarrow -2x \geq 2$

   Divide both sides by -2. Remember that you need to flip the direction of inequality sign.

   $$-2x \geq 2 \rightarrow \frac{-2x}{-2} \leq \frac{2}{-2} \rightarrow x \leq -1$$

Chapter 7: Practices

✍ **Solve each equation. (One–Step Equations)**

1) $x + 7 = 6, x =$ _____

2) $8 = 2 - x, x =$ _____

3) $-10 = 8 + x, x =$ _____

4) $x - 5 = -1, x =$ _____

5) $16 = x + 9, x =$ _____

6) $12 - x = -5, x =$ _____

✍ **Solve each equation. (Multi–Step Equations)**

7) $5(x + 3) = 20$

8) $-4(7 - x) = 16$

9) $8 = -2 (x + 5)$

10) $14 = 3(4 - 2x)$

11) $5(x + 7) = -10$

12) $-2(6 + 3x) = 12$

✍ **Solve each system of equations.**

13) $-5x + y = -3$ $x =$
     $3x - 8y = 24$ $y =$

14) $3x - 2y = 2$ $x =$
     $x - y = 2$ $y =$

15) $4x + 7y = 2$ $x =$
     $6x + 7y = 10$ $y =$

16) $5x + 7y = 18$ $x =$
     $-3x + 7y = -22$ $y =$

✍ **Draw a graph for each inequality.**

17) $x \leq -2$

18) $x > -6$

✍ **Solve each inequality and graph it.**

19) $x - 3 \geq -1$

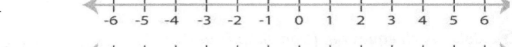

20) $3x - 2 < 16$

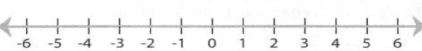

✍ **Solve each inequality.**

21) $3x + 15 > -6$

22) $-18 + 4x \leq 10$

23) $4(x + 5) \geq 8$

24) $7x - 16 < 12$

25) $3(9 + x) \geq 15$

26) $-8 + 6x > 22$

Answers – Chapter 7

1) -1

2) -6

3) -18

4) 4

5) 7

6) 17

7) 1

8) 11

9) -9

10) $-\frac{1}{3}$

11) -9

12) -4

13) $x = 0$, $y = -3$

14) $x = -2$, $y = -4$

15) $x = 4$, $y = -2$

16) $x = 5$, $y = -1$

17)

18)

19)

20)

21) $x > -7$

22) $x \le 7$

23) $x \ge -3$

24) $x < 4$

25) $x \ge -4$

26) $x > 5$

Chapter 8:

Exponents and Variables

Math Topics that you'll learn in this Chapter:

- ✓ Multiplication Property of Exponents

- ✓ Division Property of Exponents

- ✓ Powers of Products and Quotients

- ✓ Zero and Negative Exponents

- ✓ Negative Exponents and Negative Bases

- ✓ Scientific Notation

- ✓ Radicals

Multiplication Property of Exponents

☑ Exponents are shorthand for repeated multiplication of the same number by itself. For example, instead of 2×2, we can write 2^2. For $3 \times 3 \times 3 \times 3$, we can write 3^4

☑ In algebra, a variable is a letter used to stand for a number. The most common letters are: x, y, z, a, b, c, m, and n.

☑ Exponent's rules: $x^a \times x^b = x^{a+b}$, $\frac{x^a}{x^b} = x^{a-b}$

$$(x^a)^b = x^{a \times b} \qquad (xy)^a = x^a \times y^a \qquad (\frac{a}{b})^c = \frac{a^c}{b^c}$$

Examples:

1) *Multiply.* $4x^3 \times 2x^2$

   Solution: Use Exponent's rules: $x^a \times x^b = x^{a+b} \rightarrow x^3 \times x^2 = x^{3+2} = x^5$

   Then: $4x^3 \times 2x^2 = 8x^5$

2) *Simplify.* $(x^3 y^5)^2$

   Solution: Use Exponent's rules: $(x^a)^b = x^{a \times b}$. Then: $(x^3 y^5)^2 = x^{3 \times 2} y^{5 \times 2} = x^6 y^{10}$

3) *Multiply.* $-2x^5 \times 7x^3$

   Solution: Use Exponent's rules: $x^a \times x^b = x^{a+b} \rightarrow x^5 \times x^3 = x^{5+3} = x^8$

   Then: $-2x^5 \times 7x^3 = -14x^8$

4) *Simplify.* $(x^2 y^4)^3$

   Solution: Use Exponent's rules: $(x^a)^b = x^{a \times b}$. Then: $(x^2 y^4)^3 = x^{2 \times 3} y^{4 \times 3} = x^6 y^{12}$

Division Property of Exponents

☑ Exponents are shorthand for repeated multiplication of the same number by itself. For example, instead of 3×3, we can write 3^2. For $2 \times 2 \times 2$, we can write 2^3

☑ For division of exponents use following formulas:

$$\frac{x^a}{x^b} = x^{a-b} \,,\, x \neq 0, \qquad \frac{x^a}{x^b} = \frac{1}{x^{b-a}} \,,\, x \neq 0, \qquad \frac{1}{x^b} = x^{-b}$$

Examples:

1) *Simplify.* $\frac{12x^2y}{4xy^3} =$

Solution: First cancel the common factor: $4 \rightarrow \frac{12x^2y}{4xy^3} = \frac{3x^2y}{xy^3}$

Use Exponent's rules: $\frac{x^a}{x^b} = x^{a-b} \rightarrow \frac{x^2}{x} = x^{2-1} = x$ and $\frac{y}{y^3} = \frac{1}{y^{3-1}} = \frac{1}{y^2}$

Then: $\frac{12x^2y}{4xy^3} = \frac{3x}{y^2}$

2) *Simplify.* $\frac{18x^6}{2x^3} =$

Solution: Use Exponent's rules: $\frac{x^a}{x^b} = x^{b-a} \rightarrow \frac{x^6}{x^3} = x^{6-3} = x^3$

Then: $\frac{18x^6}{2x^3} = 9x^3$

3) *Simplify.* $\frac{8x^3y}{40x^2y^3} =$

Solution: First cancel the common factor: $8 \rightarrow \frac{8x^3y}{40x^2y^3} = \frac{x^3y}{5x^2y^3}$

Use Exponent's rules: $\frac{x^a}{x^b} = x^{a-b} \rightarrow \frac{x^3}{x^2} = x^{3-2} = x$

Then: $\frac{8x^3y}{40x^2y^3} = \frac{xy}{5y^3} \rightarrow$ now cancel the common factor: $y \rightarrow \frac{xy}{5y^3} = \frac{x}{5y^2}$

Powers of Products and Quotients

☑ Exponents are shorthand for repeated multiplication of the same number by itself. For example, instead of $2 \times 2 \times 2$, we can write 2^3. For $3 \times 3 \times 3 \times 3$, we can write 3^4

☑ For any nonzero numbers a and b and any integer x, $(ab)^x = a^x \times b^x$ and $(\frac{a}{b})^c = \frac{a^c}{b^c}$

Examples:

1) *Simplify.* $(6x^2y^4)^2$

   Solution: Use Exponent's rules: $(x^a)^b = x^{a \times b}$

   $(6x^2y^4)^2 = (6)^2(x^2)^2(y^4)^2 = 36x^{2 \times 2}y^{4 \times 2} = 36x^4y^8$

2) *Simplify.* $(\frac{5x}{2x^2})^2$

   Solution: First cancel the common factor: $x \rightarrow (\frac{5x}{2x^2})^2 = (\frac{5}{2x})^2$

   Use Exponent's rules: $(\frac{a}{b})^c = \frac{a^c}{b^c}$, Then: $(\frac{5}{2x})^2 = \frac{5^2}{(2x)^2} = \frac{25}{4x^2}$

3) *Simplify.* $(3x^5y^4)^2$

   Solution: Use Exponent's rules: $(x^a)^b = x^{a \times b}$

   $(3x^5y^4)^2 = (3)^2(x^5)^2(y^4)^2 = 9x^{5 \times 2}y^{4 \times 2} = 9x^{10}y^8$

4) *Simplify.* $(\frac{2x}{3x^2})^2$

   Solution: First cancel the common factor: $x \rightarrow (\frac{2x}{3x^2})^2 = (\frac{2}{3x})^2$

   Use Exponent's rules: $(\frac{a}{b})^c = \frac{a^c}{b^c}$, Then: $(\frac{2}{3x})^2 = \frac{2^2}{(3x)^2} = \frac{4}{9x^2}$

Zero and Negative Exponents

☑ Zero-Exponent Rule: $a^0 = 1$, this means that anything raised to the zero power is 1. For example: $(5xy)^0 = 1$

☑ A negative exponent simply means that the base is on the wrong side of the fraction line, so you need to flip the base to the other side. For instance, "x^{-2}" (pronounced as "ecks to the minus two") just means "x^2" but underneath, as in $\frac{1}{x^2}$.

Examples:

1) Evaluate. $\left(\frac{2}{3}\right)^{-2} =$

Solution: Use negative exponent's rule: $\left(\frac{x^a}{x^b}\right)^{-2} = \left(\frac{x^b}{x^a}\right)^2 \rightarrow \left(\frac{2}{3}\right)^{-2} = \left(\frac{3}{2}\right)^2 =$

Then: $\left(\frac{3}{2}\right)^2 = \frac{3^2}{2^2} = \frac{9}{4}$

2) Evaluate. $\left(\frac{4}{5}\right)^{-3} =$

Solution: Use negative exponent's rule: $\left(\frac{x^a}{x^b}\right)^{-2} = \left(\frac{x^b}{x^a}\right)^2 \rightarrow \left(\frac{4}{5}\right)^{-3} = \left(\frac{5}{4}\right)^3 =$

Then: $\left(\frac{5}{4}\right)^3 = \frac{5^3}{4^3} = \frac{125}{64}$

3) Evaluate. $\left(\frac{x}{y}\right)^0 =$

Solution: Use zero-exponent Rule: $a^0 = 1$
Then: $\left(\frac{x}{y}\right)^0 = 1$

4) Evaluate. $\left(\frac{5}{6}\right)^{-1} =$

Solution: Use negative exponent's rule: $\left(\frac{x^a}{x^b}\right)^{-2} = \left(\frac{x^b}{x^a}\right)^2 \rightarrow \left(\frac{5}{6}\right)^{-1} = \left(\frac{6}{5}\right)^1 = \frac{6}{5}$

Negative Exponents and Negative Bases

- ☑ A negative exponent is the reciprocal of that number with a positive exponent. $(3)^{-2} = \frac{1}{3^2}$

- ☑ To simplify a negative exponent, make the power positive!

- ☑ The parenthesis is important! -5^{-2} is not the same as $(-5)^{-2}$

$$-5^{-2} = -\frac{1}{5^2} \text{ and } (-5)^{-2} = +\frac{1}{5^2}$$

Examples:

1) *Simplify.* $\left(\frac{5a}{6c}\right)^{-2} =$

Solution: Use negative exponent's rule: $\left(\frac{x^a}{x^b}\right)^{-2} = \left(\frac{x^b}{x^a}\right)^2 \rightarrow \left(\frac{5a}{6c}\right)^{-2} = \left(\frac{6c}{5a}\right)^2$

Now use exponent's rule: $\left(\frac{a}{b}\right)^c = \frac{a^c}{b^c} \rightarrow = \left(\frac{6c}{5a}\right)^2 = \frac{6^2c^2}{5^2a^2}$

Then: $\frac{6^2c^2}{5^2a^2} = \frac{36c^2}{25a^2}$

2) *Simplify.* $\left(\frac{2x}{3yz}\right)^{-3} =$

Solution: Use negative exponent's rule: $\left(\frac{x^a}{x^b}\right)^{-2} = \left(\frac{x^b}{x^a}\right)^2 \rightarrow \left(\frac{2x}{3yz}\right)^{-3} = \left(\frac{3yz}{2x}\right)^3$

Now use exponent's rule: $\left(\frac{a}{b}\right)^c = \frac{a^c}{b^c} \rightarrow \left(\frac{3yz}{2x}\right)^3 = \frac{3^3y^3z^3}{2^3x^3} = \frac{27y^3z^3}{8x^3}$

3) *Simplify.* $\left(\frac{3a}{2c}\right)^{-2} =$

Solution: Use negative exponent's rule: $\left(\frac{x^a}{x^b}\right)^{-2} = \left(\frac{x^b}{x^a}\right)^2 \rightarrow \left(\frac{3a}{2c}\right)^{-2} = \left(\frac{2c}{3a}\right)^2$

Now use exponent's rule: $\left(\frac{a}{b}\right)^c = \frac{a^c}{b^c} \rightarrow = \left(\frac{2c}{3a}\right)^2 = \frac{2^2c^2}{3^2a^2}$

Then: $\frac{2^2c^2}{3^2a^2} = \frac{4c^2}{9a^2}$

Scientific Notation

☑ Scientific notation is used to write very big or very small numbers in decimal form.

☑ In scientific notation all numbers are written in the form of:
$m \times 10^n$, where m is greater than 1 and less than 10.

☑ To convert a number from scientific notation to standard form, move the decimal point to the left (if the exponent of ten is a negative number), or to the right (if the exponent is positive).

Examples:

1) *Write* 0.00015 *in scientific notation.*

Solution: First, move the decimal point to the right so that you have a number that is between 1 and 10. That number is 1.5
Now, determine how many places the decimal moved in step 1 by the power of 10. We moved the decimal point 4 digits to the right. Then: 10^{-4} → When the decimal moved to the right, the exponent is negative. Then: $0.00015 = 1.5 \times 10^{-4}$

2) *Write* $\mathbf{9.5 \times 10^{-5}}$ *in standard notation.*

Solution: 10^{-5} → When the decimal moved to the right, the exponent is negative.
Then: $9.5 \times 10^{-5} = 0.000095$

3) *Write* $\mathbf{0.00012}$ *in scientific notation.*

Solution: First, move the decimal point to the right so that you have a number that is between 1 and 10. Then: $m = 1.2$
Now, determine how many places the decimal moved in step 1 by the power of 10.
10^{-4} → Then: $0.00012 = 1.2 \times 10^{-4}$

4) *Write* $\mathbf{8.3 \times 10^5}$ *in standard notation.*

Solution: 10^{-5} → The exponent is positive 5. Then, move the decimal point to the right five digits. (remember $8.3 = 8.30000$)
Then: $8.3 \times 10^5 = 830000$

Radicals

☑ If n is a positive integer and x is a real number, then: $\sqrt[n]{x} = x^{\frac{1}{n}}$,

$\sqrt[n]{xy} = x^{\frac{1}{n}} \times y^{\frac{1}{n}}$, $\sqrt[n]{\frac{x}{y}} = \frac{x^{\frac{1}{n}}}{y^{\frac{1}{n}}}$, and $\sqrt[n]{x} \times \sqrt[n]{y} = \sqrt[n]{xy}$

☑ A square root of x is a number r whose square is: $r^2 = x$ (r is a square root of x.

☑ To add and subtract radicals, we need to have the same values under the radical. For example: $\sqrt{3} + \sqrt{3} = 2\sqrt{3}$, $3\sqrt{5} - \sqrt{5} = 2\sqrt{5}$

Examples:

1) *Find the square root of* $\sqrt{169}$.

   Solution: First factor the number: $169 = 13^2$,

   Then: $\sqrt{169} = \sqrt{13^2}$

   Now use radical rule: $\sqrt[n]{a^n} = a$.

   Then: $\sqrt{169} = \sqrt{13^2} = 13$

2) *Evaluate.* $\sqrt{9} \times \sqrt{25} =$

   Solution: Find the values of $\sqrt{9}$ and $\sqrt{25}$.

   Then: $\sqrt{9} \times \sqrt{25} = 3 \times 5 = 15$

3) *Solve.* $7\sqrt{2} + 4\sqrt{2}$.

   Solution: Since we have the same values under the radical, we can add these two radicals: $7\sqrt{2} + 4\sqrt{2} = 11\sqrt{2}$

4) *Evaluate.* $\sqrt{2} \times \sqrt{8} =$

   Solution: Use this radical rule: $\sqrt[n]{x} \times \sqrt[n]{y} = \sqrt[n]{xy} \rightarrow \sqrt{2} \times \sqrt{8} = \sqrt{16}$

   The square root of 16 is 4. Then: $\sqrt{2} \times \sqrt{8} = \sqrt{16} = 4$

Chapter 8: Practices

✎ *Simplify and write the answer in exponential form.*

1) $3x^3 \times 5xy^2 =$

2) $4x^2y \times 6x^2y^2 =$

3) $8x^3y^2 \times 2x^2y^3 =$

4) $7xy^4 \times 3x^2y =$

5) $6x^4y^5 \times 8x^3y^2 =$

6) $5x^3y^3 \times 8x^3y^3 =$

✎ *Simplify. (Division Property of Exponents)*

7) $\frac{5^5 \times 5^3}{5^9 \times 5} =$

8) $\frac{8x}{24x^2} =$

9) $\frac{15^4}{9x^3} =$

10) $\frac{36x^3}{54x^3y^2} =$

11) $\frac{14^3}{49^4y^4} =$

12) $\frac{120x^3y^5}{30x^2y^3} =$

✎ *Simplify. (Powers of Products and Quotients)*

13) $(8x^4y^6)^3 =$

14) $(3x^5y^4)^6 =$

15) $(5x \times 4xy^2)^2 =$

16) $\left(\frac{6x}{x^3}\right)^2 =$

17) $\left(\frac{2x^3y^5}{6x^4y^2}\right)^2 =$

18) $\left(\frac{42x^4y^6}{21x^3y^5}\right)^3 =$

✎ *Evaluate the following expressions. (Zero and Negative Exponents)*

19) $\left(\frac{2}{5}\right)^{-2} =$

20) $\left(\frac{1}{2}\right)^{-8} =$

21) $\left(\frac{2}{5}\right)^{-3} =$

22) $\left(\frac{3}{7}\right)^{-2} =$

23) $\left(\frac{5}{6}\right)^{-3} =$

24) $\left(\frac{4}{9}\right)^{-2} =$

✎ *Simplify. (Negative Exponents and Negative Bases)*

25) $16x^{-3}y^{-4} =$

26) $-9x^2y^{-3} =$

27) $12a^{-4}b^2 =$

28) $25a^3b^{-5}c^{-1} =$

29) $\frac{18y}{x^2y^{-2}} =$

30) $\frac{21a^{-2}b}{-14c^{-4}} =$

✎ *Write each number in scientific notation.*

31) $0.00615 =$

32) $0.000048 =$

33) $36,000 =$

34) $82,000,000 =$

✎ *Evaluate.*

35) $\sqrt{7} \times \sqrt{7} =$ _____

36) $\sqrt{36} - \sqrt{9} =$ _____

37) $\sqrt{25} + \sqrt{49} =$ _____

38) $\sqrt{16} \times \sqrt{64} =$ _____

39) $\sqrt{3} \times \sqrt{12} =$ _____

40) $2\sqrt{6} + 3\sqrt{6} =$ _____

Answers – Chapter 8

1) $15x^4y^2$

2) $24x^4y^3$

3) $16x^5y^5$

4) $21x^3y^5$

5) $48x^7y^7$

6) $40x^6y^6$

7) $\dfrac{1}{25}$

8) $\dfrac{1}{3x}$

9) $\dfrac{5}{3}x$

10) $\dfrac{2}{3y^3}$

11) $\dfrac{2}{7x^4y}$

12) $4xy^2$

13) $512x^{12}y^{18}$

14) $729x^{30}y^{24}$

15) $400x^4y^4$

16) $\dfrac{36}{x^4}$

17) $\dfrac{y^6}{9x^2}$

18) $8x^3y^3$

19) $\dfrac{25}{4}$

20) 256

21) $\dfrac{125}{8}$

22) $\dfrac{49}{9}$

23) $\dfrac{216}{125}$

24) $\dfrac{81}{16}$

25) $\dfrac{16}{x^3\,y^4}$

26) $-\dfrac{9x^2}{y^3}$

27) $\dfrac{12b^2}{a^4}$

28) $\dfrac{25a^3}{b^5c}$

29) $\dfrac{18y^3}{x^2}$

30) $-\dfrac{3bc^4}{2a^2}$

31) 6.15×10^{-3}

32) 4.8×10^{-5}

33) 3.6×10^4

34) 8.2×10^7

35) 7

36) 3

37) 12

38) 32

39) 6

40) $5\sqrt{6}$

Chapter 9:

Geometry and Solid Figures

Math Topics that you'll learn in this Chapter:

- ✓ The Pythagorean Theorem
- ✓ Triangles
- ✓ Polygons
- ✓ Circles
- ✓ Trapezoids
- ✓ Cubes
- ✓ Rectangle Prisms
- ✓ Cylinder

The Pythagorean Theorem

✓ You can use the Pythagorean Theorem to find a missing side in a right triangle.

✓ In any right triangle: $a^2 + b^2 = c^2$

Examples:

1) Right triangle ABC (not shown) has two legs of lengths 6 cm (AB) and 8 cm (AC). What is the length of the hypotenuse of the triangle (side BC)?

   Solution: Use Pythagorean Theorem: $a^2 + b^2 = c^2$, $a = 6$, and $b = 8$

   Then: $a^2 + b^2 = c^2 \rightarrow 6^2 + 8^2 = c^2 \rightarrow 36 + 64 = c^2 \rightarrow 100 = c^2 \rightarrow c = \sqrt{100} = 10$

   The length of the hypotenuse is 10 cm.

2) Find the hypotenuse of the following triangle.

   Solution: Use Pythagorean Theorem: $a^2 + b^2 = c^2$

   Then: $a^2 + b^2 = c^2 \rightarrow 12^2 + 5^2 = c^2 \rightarrow 144 + 25 = c^2$

   $c^2 = 169 \rightarrow c = \sqrt{169} = 13$

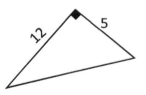

3) Find the length of the missing side in the following triangle.

   Solution: Use Pythagorean Theorem: $a^2 + b^2 = c^2$

   Then: $a^2 + b^2 = c^2 \rightarrow 3^2 + b^2 = 5^2 \rightarrow 9 + b^2 = 25 \rightarrow$

   $b^2 = 25 - 9 \rightarrow b^2 = 16 \rightarrow b = \sqrt{16} = 4$

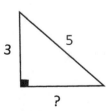

Triangles

☑ In any triangle the sum of all angles is 180 degrees.

☑ Area of a triangle $= \frac{1}{2}$ $(base \times height)$

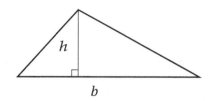

Examples:

What is the area of following triangles?

1)

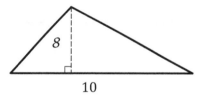

Solution:
Use the area formula: Area $= \frac{1}{2}$ $(base \times height)$
$base = 10$ and $height = 8$
Area $= \frac{1}{2}(10 \times 8) = \frac{1}{2}(80) = 40$

2)

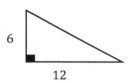

Solution:
Use the area formula: Area $= \frac{1}{2}$ $(base \times height)$
$base = 12$ and $height = 6$
Area $= \frac{1}{2}(12 \times 6) = \frac{72}{2} = 36$

3) What is the missing angle in the following triangle?

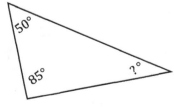

Solution:
In any triangle the sum of all angles is 180 degrees.
Let x be the missing angle. Then: $50 + 85 + x = 180$
$\rightarrow 135 + x = 180 \rightarrow x = 180 - 135 = 45$
The missing angle is 45 degrees.

Polygons

✅ Perimeter of a square

$= 4 \times side = 4s$

✅ Perimeter of a rectangle

$= 2(width + length)$

✅ Perimeter of trapezoid

$= a + b + c + d$

✅ Perimeter of a regular hexagon $= 6a$

✅ Perimeter of a parallelogram $= 2(l + w)$

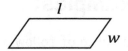

Examples:

1) Find the perimeter of following regular hexagon.

   Solution: Since the hexagon is regular, all sides are equal.

   Then: Perimeter of Hexagon $= 6 \times (one\ side)$

   Perimeter of Hexagon $= 6 \times (one\ side) = 6 \times 4 = 24\ m$

   4 m

2) Find the perimeter of following trapezoid.

   Solution: Perimeter of a trapezoid $= a + b + c + d$

   Perimeter of the trapezoid $= 5 + 6 + 6 + 8 = 25\ ft$

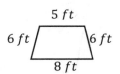

Circles

☑ In a circle, variable r is usually used for the radius and d for diameter.

☑ *Area of a circle* $= \pi r^2$ (π is about 3.14)

☑ *Circumference of a circle* $= 2\pi r$

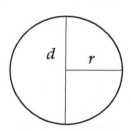

Examples:

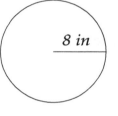

1) Find the area of the following circle.

Solution:

Use area formula: $Area = \pi r^2$

$r = 8 \, in \rightarrow Area = \pi(8)^2 = 64\pi, \pi = 3.14$

Then: $Area = 64 \times 3.14 = 200.96 \, in^2$

2) Find the Circumference of the following circle.

Solution:

Use Circumference formula: $Circumference = 2\pi r$

$r = 5 \, cm \rightarrow Circumference = 2\pi(5) = 10\pi$

$\pi = 3.14$ **Then:** $Circumference = 10 \times 3.14 = 31.4 \, cm$

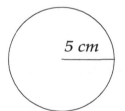

3) Find the area of the circle.

Solution:

Use area formula: $Area = \pi r^2$,

$r = 5 \, in$ then: $Area = \pi(5)^2 = 25\pi, \pi = 3.14$

Then: $Area = 25 \times 3.14 = 78.5$

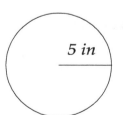

Trapezoids

☑ A quadrilateral with at least one pair of parallel sides is a trapezoid.

☑ Area of a trapezoid $= \frac{1}{2}h(b_1 + b_2)$

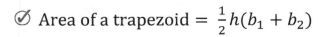

Examples:

1) Calculate the area of the following trapezoid.

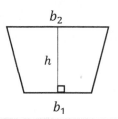

Solution:

Use area formula: $A = \frac{1}{2}h(b_1 + b_2)$

$b_1 = 5\ cm$, $b_2 = 8\ cm$ and $h = 10\ cm$

Then: $A = \frac{1}{2}(10)(8 + 5) = 5(13) = 65\ cm^2$

2) Calculate the area of the following trapezoid.

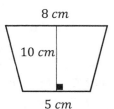

Solution:

Use area formula: $A = \frac{1}{2}h(b_1 + b_2)$

$b_1 = 12\ cm$, $b_2 = 20\ cm$ and $h = 14\ cm$

Then: $A = \frac{1}{2}(14)(12 + 20) = 7(32) = 224\ cm^2$

Cubes

☑ A cube is a three-dimensional solid object bounded by six square sides.

☑ Volume is the measure of the amount of space inside of a solid figure, like a cube, ball, cylinder or pyramid.

☑ Volume of a cube = $(one\ side)^3$

☑ surface area of a cube = $6 \times (one\ side)^2$

Examples:

1) Find the volume and surface area of the following cube.

Solution: Use volume formula: $volume = (one\ side)^3$

Then: $volume = (one\ side)^3 = (2)^3 = 8\ cm^3$

Use surface area formula: $surface\ area\ of\ cube: 6(one\ side)^2 =$
$$6(2)^2 = 6(4) = 24\ cm^2$$

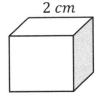

2) Find the volume and surface area of the following cube.

Solution: Use volume formula: $volume = (one\ side)^3$

Then: $volume = (one\ side)^3 = (5)^3 = 125\ cm^3$

Use surface area formula:

$surface\ area\ of\ cube: 6(one\ side)^2 = 6(5)^2 = 6(25) = 150\ cm^2$

3) Find the volume and surface area of the following cube.

Solution: Use volume formula: $volume = (one\ side)^3$

Then: $volume = (one\ side)^3 = (7)^3 = 343\ m^3$

Use surface area formula:

$surface\ area\ of\ cube: 6(one\ side)^2 = 6(7)^2 = 6(49) = 294\ m^2$

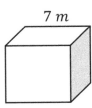

Rectangular Prisms

✅ A rectangular prism is a solid 3-dimensional object which has six rectangular faces.

✅ Volume of a Rectangular prism $= \boldsymbol{Length \times Width \times Height}$

$Volume = l \times w \times h$

$Surface\ area = 2 \times (wh + lw + lh)$

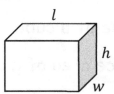

Examples:

1) Find the volume and surface area of the following rectangular prism.

   Solution:

   Use volume formula: $Volume = l \times w \times h$

   Then: $Volume = 8 \times 6 \times 10 = 480\ m^3$

   Use surface area formula: $Surface\ area = 2 \times (wh + lw + lh)$

   Then: $Surface\ area = 2 \times \big((6 \times 10) + (8 \times 6) + (8 \times 10)\big)$

   $\qquad\qquad = 2 \times (60 + 48 + 80) = 2 \times (188) = 376\ m^2$

2) Find the volume and surface area of rectangular prism.

   Solution:

   Use volume formula: $Volume = l \times w \times h$

   Then: $Volume = 10 \times 8 \times 12 = 960\ m^3$

   Use surface area formula: $Surface\ area = 2 \times (wh + lw + lh)$

   Then: $Surface\ area = 2 \times \big((8 \times 12) + (10 \times 8) + (10 \times 12)\big)$

   $\qquad\qquad = 2 \times (96 + 80 + 120) = 2 \times (296) = 592\ m^2$

Cylinder

☑ A cylinder is a solid geometric figure with straight parallel sides and a circular or oval cross section.

☑ *Volume of a Cylinder* $= \pi(radius)^2 \times height, \pi \approx 3.14$

☑ *Surface area of a cylinder* $= 2\pi r^2 + 2\pi rh$

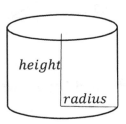

Examples:

1) *Find the volume and Surface area of the follow Cylinder.*

Solution:

Use volume formula: $Volume = \pi(radius)^2 \times height$
Then: $Volume = \pi(3)^2 \times 8 = 9\pi \times 8 = 72\pi$
$\pi = 3.14$ **then:** $Volume = 72\pi = 72 \times 3.14 = 226.08 \; cm^3$
Use surface area formula: $Surface \; area = 2\pi r^2 + 2\pi rh$
Then: $2\pi(3)^2 + 2\pi(3)(8) = 2\pi(9) + 2\pi(24) = 18\pi + 48\pi = 66\pi$
$\pi = 3.14$ Then: $Surface \; area = 66 \times 3.14 = 207.24 \; cm^2$

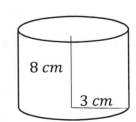

2) *Find the volume and Surface area of the follow Cylinder.*

Solution:

Use volume formula: $Volume = \pi(radius)^2 \times height$
Then: $Volume = \pi(2)^2 \times 6 = \pi4 \times 6 = 24\pi$
$\pi = 3.14$ **then:** $Volume = 24\pi = 75.36 \; cm^3$
Use surface area formula: $Surface \; area = 2\pi r^2 + 2\pi rh$
Then: $= 2\pi(2)^2 + 2\pi(2)(6) = 2\pi(4) + 2\pi(12) = 8\pi + 24\pi = 32\pi$
$\pi = 3.14$ **then:** $Surface \; area = 32 \times 3.14 = 100.48 \; cm^2$

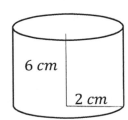

Chapter 9: Practices

 Find the missing side?

1)

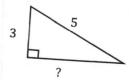
3 5 ?

2)

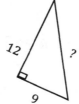
12 ? 9

3)

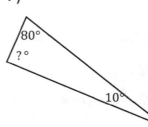
20 ? 16

4)

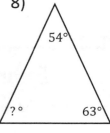
5 ? 12

 Find the measure of the unknown angle in each triangle.

5)

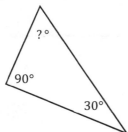
?° 90° 30°

6)

7)

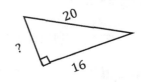
80° ?° 10°

8)

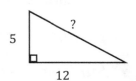
54° ?° 63°

 Find area of each triangle.

9)   10)   11)   12)

 Find the perimeter of each shape.

13)   14)   15)   16)

86

✏ **Complete the table below.** (π = 3.14)

17)

	Radius	Diameter	Circumference	Area
Circle 1	3 *inches*	6 *inches*	18.84 *inches*	28.26 *square inches*
Circle 2			43.96 *meters*	
Circle 3		8 *ft*		
Circle 4				78.5 *square miles*

✏ **Find the area of each trapezoid.**

18)   19)   20)   21)

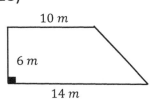
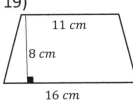
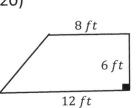
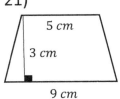

✏ **Find the volume of each cube.**

22)   23)   24)   25)

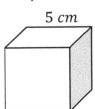
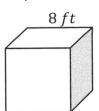
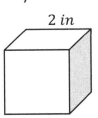
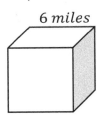

✏ **Find the volume of each Rectangular Prism.**

26)   27)   28)

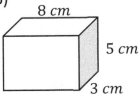
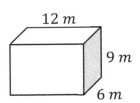
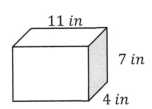

✍ Find the volume of each Cylinder. Round your answer to the nearest tenth. (π = 3.14)

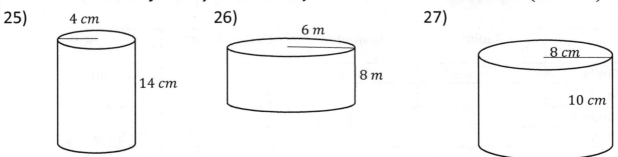

25)   4 cm

14 cm

26)   6 m

8 m

27)   8 cm

10 cm

Answers – Chapter 9

1) 4

2) 15

3) 12

4) 13

5) 60°

6) 80°

7) 90°

8) 63°

9) 24 *square unites*

10) 30 *square unites*

11) 64 cm^2

12) 20 in^2

13) 64 cm

14) 30 ft

15) 40 in

16) 30 m

17)

	Radius	Diameter	Circumference	Area
Circle 1	3 inches	6 inches	18.84 inches	28.26 square inches
Circle 2	7 meters	14 meters	43.96 meters	153.86 square meters
Circle 3	4 ft	8 ft	25.12 ft	50.24 square ft
Circle 4	5 miles	10 miles	31.4 miles	78.5 square miles

18) 72 m^2

19) 108 cm^2

20) 60 ft^2

21) 21 cm^2

22) 125 cm^3

23) 512 ft^3

24) 8 in^3

25) 216 $miles^3$

26) 120 cm^3

27) 648 m^3

28) 308 in^3

29) 703.36 cm^3

30) 904.32 m^3

31) 2,009.6 cm^3

Chapter 10:

Statistics

Math Topics that you'll learn in this Chapter:

- ✓ Mean, Median, Mode, and Range of the Given Data

- ✓ Pie Graph

- ✓ Probability Problems

- ✓ Permutations and Combinations

Mean, Median, Mode, and Range of the Given Data

☑ Mean: $\dfrac{sum\ of\ the\ data}{total\ number\ of\ data\ entires}$

☑ Mode: the value in the list that appears most often

☑ Median: is the middle number of a group of numbers that have been arranged in order by size.

☑ Range: the difference of largest value and smallest value in the list

Examples:

1) **What is the mode of these numbers?** $4, 5, 7, 5, 7, 4, 0, 4$

   Solution: Mode: the value in the list that appears most often.
   Therefore, the mode is number 4. There are three number 4 in the data.

2) **What is the median of these numbers?** $5, 10, 14, 9, 16, 19, 6$

   Solution: Write the numbers in order: $5, 6, 9, 10, 14, 16, 19$

   Median is the number in the middle. Therefore, the median is 10.

3) **What is the mean of these numbers?** $8, 2, 8, 5, 3, 2, 4, 8$

   Solution: Mean: $\dfrac{sum\ of\ the\ data}{total\ number\ of\ data\ entires} = \dfrac{8+2+8+5+3+2+4+8}{8} = 5$

4) **What is the range in this list?** $4, 9, 13, 8, 15, 18, 5$

   Solution: Range is the difference of largest value and smallest value in the list. The largest value is 18 and the smallest value is 4. Then: $18 - 4 = 14$

Pie Graph

☑ A Pie Chart is a circle chart divided into sectors, each sector represents the relative size of each value.

☑ Pie charts represent a snapshot of how a group is broken down into smaller pieces.

Example:

A library has 820 books that include Mathematics, Physics, Chemistry, English and History. Use following graph to answer the questions.

1) What is the number of Mathematics books?

Solution: Number of total books = 820

Percent of Mathematics books = 30% = 0.30

Then, number of Mathematics books:
$$0.30 \times 820 = 246$$

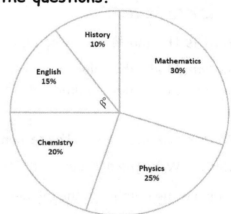

2) What is the number of History books?

Solution: Number of total books = 820

Percent of History books = 10% = 0.10

Then: $0.10 \times 820 = 82$

3) What is the number of Chemistry books?

Solution: Number of total books = 820

Percent of Chemistry books = 20% = 0.20

Then: $0.20 \times 820 = 164$

Probability Problems

✅ Probability is the likelihood of something happening in the future. It is expressed as a number between zero (can never happen) to 1 (will always happen).

✅ Probability can be expressed as a fraction, a decimal, or a percent.

✅ Probability formula: $Probability = \frac{number\ of\ desired\ outcomes}{number\ of\ total\ outcomes}$

Examples:

1) Anita's trick-or-treat bag contains 12 pieces of chocolate, 18 suckers, 18 pieces of gum, 24 pieces of licorice. If she randomly pulls a piece of candy from her bag, what is the probability of her pulling out a piece of sucker?

Solution: $Probability = \frac{number\ of\ desired\ outcomes}{number\ of\ total\ outcomes}$

Probability of ulling out a piece of sucker $= \frac{18}{12+18+18+24} = \frac{18}{72} = \frac{1}{4}$

2) A bag contains 20 balls: four green, five black, eight blue, a brown, a red and one white. If 19 balls are removed from the bag at random, what is the probability that a brown ball has been removed?

Solution: If 19 balls are removed from the bag at random, there will be one ball in the bag. The probability of choosing a brown ball is 1 out of 20. Therefore, the probability of not choosing a brown ball is 19 out of 20 and the probability of having not a brown ball after removing 19 balls is the same.

Permutations and Combinations

☑ Factorials are products, indicated by an exclamation mark. For example, $4! = 4 \times 3 \times 2 \times 1$(Remember that 0! is defined to be equal to 1.)

☑ Permutations: The number of ways to choose a sample of k elements from a set of n distinct objects where order does matter, and replacements are not allowed. For a permutation problem, use this formula:

$$_nP_k = \frac{n!}{(n-k)!}$$

☑ Combination: The number of ways to choose a sample of r elements from a set of n distinct objects where order does not matter, and replacements are not allowed. For a combination problem, use this formula:

$$_nC_r = \frac{n!}{r!\,(n-r)!}$$

Examples:

1) *How many ways can the first and second place be awarded to 8 people?*

   Solution: Since the order matters, (the first and second place are different!) we need to use permutation formula where n is 10 and k is 2. Then: $\frac{n!}{(n-k)!} = \frac{8!}{(8-2)!} = \frac{8!}{6!} = \frac{8 \times 7 \times 6!}{6!}$, remove 6! from both sides of the fraction. Then: $\frac{8 \times 7 \times 6!}{6!} = 8 \times 7 = 56$

2) *How many ways can we pick a team of 2 people from a group of 6?*

   Solution: Since the order doesn't matter, we need to use combination formula where n is 8 and r is 3. Then: $\frac{n!}{r!\,(n-r)!} = \frac{6!}{2!\,(6-2)!} = \frac{6!}{2!\,(4)!} = \frac{6 \times 5 \times 4!}{2!\,(4)!} = \frac{6 \times 5}{2 \times 1} = \frac{30}{2} = 15$

Chapter 10: Practices

✍ *Find the values of the Given Data.*

1) 6, 12, 1, 1, 5

   Mode: _____   Range: _____

   Mean: _____   Median: _____

2) 5, 8, 3, 7, 4, 3

   Mode: _____   Range: _____

   Mean: _____   Median: _____

✍   The circle graph below shows all Jason's expenses for last month. Jason spent $864 on his bills last month.

3) How much did Jason spend on his car last month? _____

4) How much did Jason spend for foods last month? _____

Jason's last month expenses

✎ Solve.

5) Bag A contains 8 red marbles and 6 green marbles. Bag B contains 5 black marbles and 10 orange marbles. What is the probability of selecting a green marble at random from bag A? What is the probability of selecting a black marble at random from Bag B? _____   _____

✎ Solve.

6) Susan is baking cookies. She uses sugar, flour, butter, and eggs. How many different orders of ingredients can she try? _____

7) Jason is planning for his vacation. He wants to go to museum, watch a movie, go to the beach, and play volleyball. How many different ways of ordering are there for him? _____

8) In how many ways a team of 10 basketball players can to choose a captain and co-captain? _____

9) How many ways can you give 3 balls to your 8 friends? _____

10) A professor is going to arrange her 8 students in a straight line. In how many ways can she do this? _____

Answers – Chapter 10

1) Mode: 1 , Range: 11, Mean: 5, Median: 5

2) Mode: 3 , Range: 5 , Mean: 5, Median: 4.5

3) $640

4) $448

5) $\frac{3}{7}, \frac{1}{3}$

6) 24

7) 24

8) 90

9) 56

10) 40,320

ISEE Middle Level Test Review

The Independent School Entrance Exam (ISEE) is an admission test developed by the Educational Records Bureau for its member schools as part of their admission process.

ISEE Middle Level tests use a multiple-choice format and contain two Mathematics sections:

Quantitative Reasoning

There are 37 questions in the Quantitative Reasoning section and students have 35 minutes to answer the questions. This section contains word problems and quantitative comparisons. The word problems require either no calculation or simple calculation. The quantitative comparison items present two quantities, (A) and (B), and the student needs to select one of the following four answer choices:

(A) The quantity in Column A is greater.

(B) The quantity in Column B is greater.

(C) The two quantities are equal.

(D) The relationship cannot be determined from the information given.

Mathematics Achievement

There are 47 questions in the Mathematics Achievement section and students have 40 minutes to answer the questions. Mathematics Achievement measures students' knowledge of Mathematics requiring one or more steps in calculating the answer.

In this book, there are two complete ISEE Middle Level Quantitative Reasoning and Mathematics Achievement practice tests. Take these tests to see what score you'll be able to receive on a real ISEE Lower Level test.

Good luck!

Time to refine your skill with a practice examination

Take practice ISEE Middle Level Math Tests to simulate the test day experience. After you've finished, score your tests using the answer keys.

Before You Start

- You'll need a pencil and a timer to take the test.

- After you've finished the test, review the answer key to see where you went wrong.

- Use the answer sheet provided to record your answers. (You can cut it out or photocopy it)

- Students receive 1 point for every correct answer. There is no penalty for wrong or skipped questions.

Calculators are NOT permitted for the ISEE Middle Level Test

Good Luck!

ISEE Middle Level Math

Practice Test 1

2020 - 2021

Two Parts

Total number of questions: 84

Part 1 (Quantitative Reasoning): 37 questions

Part 2 (Mathematics Achievement): 47 questions

Total time for two parts: 75 Minutes

ISEE Middle Level Practice Test Answer Sheets

Remove (or photocopy) this answer sheet and use it to complete the practice test.

ISEE Middle Level Practice Test 1

Quantitative Reasoning   Mathematics Achievement

#	QR	#	QR	#	MA	#	MA
1	Ⓐ Ⓑ Ⓒ Ⓓ	25	Ⓐ Ⓑ Ⓒ Ⓓ	1	Ⓐ Ⓑ Ⓒ Ⓓ	25	Ⓐ Ⓑ Ⓒ Ⓓ
2	Ⓐ Ⓑ Ⓒ Ⓓ	26	Ⓐ Ⓑ Ⓒ Ⓓ	2	Ⓐ Ⓑ Ⓒ Ⓓ	26	Ⓐ Ⓑ Ⓒ Ⓓ
3	Ⓐ Ⓑ Ⓒ Ⓓ	27	Ⓐ Ⓑ Ⓒ Ⓓ	3	Ⓐ Ⓑ Ⓒ Ⓓ	27	Ⓐ Ⓑ Ⓒ Ⓓ
4	Ⓐ Ⓑ Ⓒ Ⓓ	28	Ⓐ Ⓑ Ⓒ Ⓓ	4	Ⓐ Ⓑ Ⓒ Ⓓ	28	Ⓐ Ⓑ Ⓒ Ⓓ
5	Ⓐ Ⓑ Ⓒ Ⓓ	29	Ⓐ Ⓑ Ⓒ Ⓓ	5	Ⓐ Ⓑ Ⓒ Ⓓ	29	Ⓐ Ⓑ Ⓒ Ⓓ
6	Ⓐ Ⓑ Ⓒ Ⓓ	30	Ⓐ Ⓑ Ⓒ Ⓓ	6	Ⓐ Ⓑ Ⓒ Ⓓ	30	Ⓐ Ⓑ Ⓒ Ⓓ
7	Ⓐ Ⓑ Ⓒ Ⓓ	31	Ⓐ Ⓑ Ⓒ Ⓓ	7	Ⓐ Ⓑ Ⓒ Ⓓ	31	Ⓐ Ⓑ Ⓒ Ⓓ
8	Ⓐ Ⓑ Ⓒ Ⓓ	32	Ⓐ Ⓑ Ⓒ Ⓓ	8	Ⓐ Ⓑ Ⓒ Ⓓ	32	Ⓐ Ⓑ Ⓒ Ⓓ
9	Ⓐ Ⓑ Ⓒ Ⓓ	33	Ⓐ Ⓑ Ⓒ Ⓓ	9	Ⓐ Ⓑ Ⓒ Ⓓ	33	Ⓐ Ⓑ Ⓒ Ⓓ
10	Ⓐ Ⓑ Ⓒ Ⓓ	34	Ⓐ Ⓑ Ⓒ Ⓓ	10	Ⓐ Ⓑ Ⓒ Ⓓ	34	Ⓐ Ⓑ Ⓒ Ⓓ
11	Ⓐ Ⓑ Ⓒ Ⓓ	35	Ⓐ Ⓑ Ⓒ Ⓓ	11	Ⓐ Ⓑ Ⓒ Ⓓ	35	Ⓐ Ⓑ Ⓒ Ⓓ
12	Ⓐ Ⓑ Ⓒ Ⓓ	36	Ⓐ Ⓑ Ⓒ Ⓓ	12	Ⓐ Ⓑ Ⓒ Ⓓ	36	Ⓐ Ⓑ Ⓒ Ⓓ
13	Ⓐ Ⓑ Ⓒ Ⓓ	37	Ⓐ Ⓑ Ⓒ Ⓓ	13	Ⓐ Ⓑ Ⓒ Ⓓ	37	Ⓐ Ⓑ Ⓒ Ⓓ
14	Ⓐ Ⓑ Ⓒ Ⓓ			14	Ⓐ Ⓑ Ⓒ Ⓓ	38	Ⓐ Ⓑ Ⓒ Ⓓ
15	Ⓐ Ⓑ Ⓒ Ⓓ			15	Ⓐ Ⓑ Ⓒ Ⓓ	39	Ⓐ Ⓑ Ⓒ Ⓓ
16	Ⓐ Ⓑ Ⓒ Ⓓ			16	Ⓐ Ⓑ Ⓒ Ⓓ	40	Ⓐ Ⓑ Ⓒ Ⓓ
17	Ⓐ Ⓑ Ⓒ Ⓓ			17	Ⓐ Ⓑ Ⓒ Ⓓ	41	Ⓐ Ⓑ Ⓒ Ⓓ
18	Ⓐ Ⓑ Ⓒ Ⓓ			18	Ⓐ Ⓑ Ⓒ Ⓓ	42	Ⓐ Ⓑ Ⓒ Ⓓ
19	Ⓐ Ⓑ Ⓒ Ⓓ			19	Ⓐ Ⓑ Ⓒ Ⓓ	43	Ⓐ Ⓑ Ⓒ Ⓓ
20	Ⓐ Ⓑ Ⓒ Ⓓ			20	Ⓐ Ⓑ Ⓒ Ⓓ	44	Ⓐ Ⓑ Ⓒ Ⓓ
21	Ⓐ Ⓑ Ⓒ Ⓓ			21	Ⓐ Ⓑ Ⓒ Ⓓ	45	Ⓐ Ⓑ Ⓒ Ⓓ
22	Ⓐ Ⓑ Ⓒ Ⓓ			22	Ⓐ Ⓑ Ⓒ Ⓓ	46	Ⓐ Ⓑ Ⓒ Ⓓ
23	Ⓐ Ⓑ Ⓒ Ⓓ			23	Ⓐ Ⓑ Ⓒ Ⓓ	47	Ⓐ Ⓑ Ⓒ Ⓓ
24	Ⓐ Ⓑ Ⓒ Ⓓ			24	Ⓐ Ⓑ Ⓒ Ⓓ		

ISEE Middle Level Math
Practice Test 1

Section 1

37 questions

Total time for this section: 35 Minutes

You may NOT use a calculator for this test.

1) Solve. $\dfrac{-50 \times 0.5}{5}$
A. -16
B. -5
C. 5
D. 16

2) A $41 shirt now selling for $29 is discounted by what percent?
A. 20%
B. 29%
C. 40%
D. 50%

3) $562{,}357{,}741 \times 0.0001$?
A. $562{,}357.741$
B. $56{,}235.7741$
C. $5{,}623.57741$
D. 562.357741

4) Jim purchased a table for 30% off and saved $24. What was the original price of the table?
(A) $80
(B) $110
(C) $115
(D) 150

5) If $f = 3x - 2y$ and $g = x + 5y$, what is $3f + g$?
A. $5x - y$
B. $5x - 2y$
C. $10x - 2y$
D. $10x - y$

6) Which of the following shows the numbers in increasing order?
A. $\dfrac{1}{7}, \dfrac{3}{5}, \dfrac{1}{3}, \dfrac{3}{4}$
B. $\dfrac{1}{7}, \dfrac{3}{5}, \dfrac{3}{4}, \dfrac{1}{3}$
C. $\dfrac{1}{7}, \dfrac{1}{3}, \dfrac{3}{5}, \dfrac{3}{4}$
D. $\dfrac{1}{7}, \dfrac{3}{4}, \dfrac{1}{3}, \dfrac{3}{5}$

7) What is the value of x in the following equation? $8^x = 512$
A. 2
B. 3
C. 4
D. 5

8) What is the value of x in the following figure?

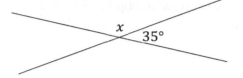

A. 35°
B. 75°
C. 115°
D. 145°

9) The score of Emma was half as that of Ava and the score of Mia was twice that of Ava. If the score of Mia was 40, what is the score of Emma?
(A) 10
(B) 18
(C) 20
(D) 30

10) The area of a circle is 49 π. What is the circumference of the circle?
A. 8π
B. 14π
C. 32π
D. 49π

11) Two third of 15 is equal to $\frac{5}{2}$ of what number?
(A) 25
(B) 15
(C) 10
(D) 4

12) What is round off the result of 1.15×8.2 to the nearest tenth?
A. 6
B. 7
C. 8.06
D. 9.4

ISEE Middle Level Math Prep 2020-2021

13) The perimeter of the trapezoid below is 54. What is its area?

(A) $255\ cm^2$
(B) $234\ cm^2$
(C) $150\ cm^2$
(D) $100\ cm^2$

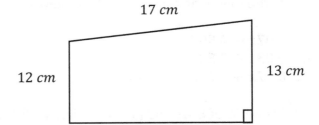

14) In five successive hours, a car traveled $42\ km, 46\ km, 52\ km, 35\ km$ and $58\ km$. In the next five hours, it traveled with an average speed of $60\ km\ per\ hour$. Find the total distance the car traveled in 10 hours.
A. $435\ km$
B. $450\ km$
C. $475\ km$
D. $533\ km$

15) What is the mean in the following set of numbers?

$$10, 13, 29, 37, 46, 66, 100, 124$$

A. 46.2
B. 40.5
C. 51.4
D. 53.12

16) Find $\frac{1}{2}$ of $\frac{2}{5}$ of 145?
A. 30
B. 29
C. 18
D. 4

17) The price of a laptop is decreased by 20% to $320. What is its original price?
A. $320
B. $380
C. $400
D. $455

18) A company pays its employee $7,500 plus 3% of all sales profit. If x is all sold profit, which of the following represents the employee's revenue?

A. $0.03x$
B. $0.97x - 7,500$
C. $0.03x + 7,500$
D. $0.97x + 7,500$

19) The ratio of boys and girls in a class is $7:4$. If there are 44 students in the class, how many more girls should be enrolled to make the ratio $1:1$?

A. 8
B. 10
C. 12
D. 16

20) Which of the following is a correct statement?

(A) $\frac{3}{5} > 0.8$

(B) $12\% = \frac{2}{5}$

(C) $3 < \frac{5}{2}$

(D) $\frac{7}{6} > 0.8$

21) What is the value of x in the following equation?
$$6(x + 1) = 4(x - 4) + 20$$

A. 12
B. -12
C. 1
D. -1

22) What is the value of x in the following figure?

A. 160
B. 145
C. 125
D. 105

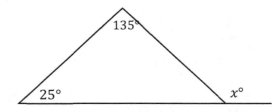

23) Car A use 4-liter petrol per 140 kilometers; car B use 3-liter petrol per 140 kilometers. If both cars drive 350 kilometers, how much more petrol does car A use?

A. 2.5
B. 10
C. 15.5
D. 25

24) What is the area of a square whose diagonal is 6?

A. 18
B. 32
C. 46
D. 64

25) What is value of $-25 - (-66)$?

A. 41
B. -91
C. -41
D. 91

Quantitative Comparisons

Direction: Questions 26 to 37 are Quantitative Comparisons Questions. Using the information provided in each question, compare the quantity in column A to the quantity in Column B. Choose on your answer sheet grid

- A if the quantity in Column A is greater
- B if the quantity in Column B is greater
- C if the two quantities are equal
- D if the relationship cannot be determined from the information given

26)

Column A	**Column B**
$\sqrt{36} + \sqrt{36}$	$\sqrt{72}$

27) $y = -4x - 8$

Column A	**Column B**
The value of x when $y = 12$	-4

28)

Column A	**Column B**
$6 + 4 \times 7 + 8$	$4 + 6 \times 7 - 8$

29) The average age of Joe, Michelle, and Nicole is 32.

Column A	**Column B**
The average age of Joe and Michelle	The average age of Michelle and Nicole

30)

Column A	**Column B**
$\sqrt{121 - 64}$	$\sqrt{121} - \sqrt{64}$

ISEE Middle Level Math Prep 2020-2021

31) A right cylinder with radius 2 inches has volume 50π cubic inches.

Quantity A	Quantity B
The height of the cylinder	10 inches

32) x is an integer.

Quantity A	Quantity B
$\frac{x^6}{6}$	$\left(\frac{x}{6}\right)^6$

33) x is an integer greater than zero.

Quantity A	Quantity B
$\frac{1}{x} + x$	8

34)

$$\frac{4}{5} < x < \frac{6}{7}$$

Quantity A	Quantity B
x	$\frac{5}{6}$

35) a and b are real numbers.

$$a < b$$

Quantity A	Quantity B
$|a - b|$	$|b - a|$

36) $2x^3 + 10 = 64$
    $120 - 18y = 84$

Quantity A	Quantity B
x	y

37) The average of $3, 4$, and x is 3.

Quantity A	Quantity B
x	average of $x, x - 6, x + 4, 2x$

ISEE Middle Level Math Prep 2020-2021

ISEE Middle Level Math
Practice Test 1

Section 2

47 questions

Total time for this section: 40 Minutes

You may NOT use a calculator for this test.

1) What number is 5 less than 50% of 46?
A. 10
B. 13
C. 18
D. 23

2) $4\left(\frac{1}{3} - \frac{1}{6}\right) + 5$?
A. 4
B. 4.5
C. 5.66 …
D. 5

3) What number is 15 more than 20% of 120?
A. 39
B. 29
C. 25
D. 20

4) In a bundle of 85 pencils, 42 are red and the rest are blue. About what percent of the bundle is composed of blue pencils?
A. 62%
B. 58%
C. 54%
D. 51%

5) What is the value of x in the following equation?
$$(x + 6)^3 = 64$$
A. 1
B. -1
C. 2
D. -2

6) If a box contains red and blue balls in ratio of $3: 2$, how many red balls are there if 90 blue balls are in the box?
A. 140
B. 135
C. 60
D. 10

7) What is the difference in perimeter between a 8 *cm* by 5 *cm* rectangle and a circle with diameter of 12 *cm*? ($\pi = 3$)
A. 8 *cm*
B. 9 *cm*
C. 10 *cm*
D. 11 *cm*

8) When a number is subtracted from 27 and the difference is divided by that number, the result is 2. What is the value of the number?
A. 2
B. 5
C. 9
D. 12

9) If $\frac{3x}{2} = 15$, then $\frac{2x}{5} = ?$
A. 4
B. 8
C. 10
D. 20

10) The price of a car was $20,000 in 2014, and it was $15,000 in 2015. What is the rate of depreciation of the price of car per year?
A. 15%
B. 25%
C. 30%
D. 35%

11) Which of the following is the greatest number?
A. $\frac{1}{4}$
B. $\frac{7}{9}$
C. 0.85
D. 75%

12) Calculate the approximate area of the following circle. (The diameter of the circle is 10 cm)
A. 1250 cm^2
B. 314cm^2
C. 116 cm^2
D. 78.5 cm^2

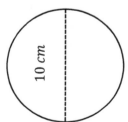

13) 90 is equal to?

A. $2 + (3 \times 10) + (2 \times 30)$

B. $\left(\frac{10}{3} \times 27\right) + \left(\frac{5}{2} \times 2\right)$

C. $\left(\left(\frac{3}{2} + 3\right) \times \frac{18}{3}\right) + 63$

D. $(2 \times 15) + (50 \times 2) - 46$

14) Which of the following angles can represent the three angles of an isosceles right triangle?

A. $45°, 90°, 45°$

B. $50°, 50°, 80°$

C. $60°, 60°, 60°$

D. $55°, 35°, 90°$

15) If 120% of a number is 84, then what is 90% of that number?

A. 45

B. 63

C. 74

D. 84

16) What is the missing term in the given numbers?

$$3, 4, 6, 9, 13, 18, 24, \underline{\hspace{1cm}}, 39$$

A. 24

B. 26

C. 27

D. 31

17) In following rectangle which statement is true?

A. Length of AB is smaller than BC.

B. The sum of all the angles equals 360 degrees.

C. Length of AB equal to length DC.

D. AB is perpendicular to DC.

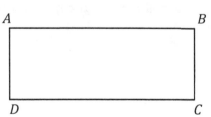

18) A football team had $25,000 to spend on supplies. The team spent $15,000 on new balls. New sport shoes cost $130 each. Which of the following inequalities represent how many new shoes the team can purchase?

A. $130x + 15,000 \leq 25,000$
B. $130x + 15,000 \geq 25,000$
C. $15,000x + 130 \leq 25,000$
D. $15,000x + 130 \geq 25,000$

19) The capacity of a red box is 30% greater than a blue box. If the capacity of the red box is 52 books, how many books can be put in the blue box?

A. 9
B. 15
C. 40
D. 42

20) From last year, the price of gasoline has increased from $1.40 per gallon to $1.75 per gallon. The new price is what percent of the original price?

A. 72%
B. 125%
C. 140%
D. 160%

21) When a gas tank can hold 30 gallons, how many gallons does it contain when it is $\frac{2}{3}$ full?

A. 115
B. 65.5
C. 20
D. 10

22) 190 minutes = ...?

A. $3.25 \; Hours$
B. $3.16 \; Hours$
C. $2 \; Hours$
D. $0.4 \; Hours$

23) Which of the following is **NOT** a prime number?

A. 107
B. 101
C. 71
D. 58

24) What is the perimeter of a square that has an area of 36 square inches?
A. 144 inches
B. 36 inches
C. 24 inches
D. 56 inches

25) Jason left a $13.00 tip on a lunch that cost $26.00, what percentage was the tip?

A. 2.5%
B. 10%
C. 25%
D. 50%

26) A box of 36 pencils costs $1.80, what is the unit cost?
A. $0.50
B. $0.25
C. $0.08
D. $0.05

27) Two-kilograms apple and two-kilograms orange cost $26.4 If one-kilogram apple costs $4.2 how much does one-kilogram orange cost?
A. $9
B. $6
C. $4.5
D. $4

28) $[6 \times (-24) + 8] - (-4) + [6 \times 5] \div 2 \ = \ ?$
A. 158
B. 132
C. −104
D. −117

29) The width of a rectangle is $3x$ and its length is $5x$. The perimeter of the rectangle is 80. What is the value of x?
A. 4
B. 5
C. 6
D. 10

30) Jason is 15 miles ahead of Joe running at 5.5 miles per hour and Joe is running at the speed of 7 miles per hour. How long does it take Joe to catch Jason?

A. 4 hours

B. 6 *hours*

C. 8*hours*

D. 10 *hours*

31) $\left(\left((-12) + 20\right) \times 3\right) + (-16)?$

A. 1

B. 4

C. 6

D. 8

32) If 45% of a class are girls, and 20% of girls play tennis, what percent of the class play tennis?

A. 9%

B. 15%

C. 25%

D. 40%

33) The price of a sofa is decreased by 30% to $490. What was its original price?

A. $480

B. $510

C. $550

D. $700

34) In a class, there are twice as many girls as boys. If the total number of students in the class is 45, how many girls are in the class?

A. 15

B. 25

C. 32

D. 35

35) At a Zoo, the ratio of lions to tigers is 5 to 3. Which of the following could NOT be the total number of lions and tigers in the zoo?

A. 32

B. 40

C. 97

D. 104

36) Solving the equation: $10x - 15.5 = -55.5$?
A. -4
B. -3
C. 4
D. 3

37) A shaft rotates 400 times in 5 seconds. How many times does it rotate in 0.25 minutes?
A. 1,200
B. 950
C. 400
D. 100

38) A swimming pool holds 2,500 cubic feet of water. The swimming pool is 40 feet long and 10 feet wide. How deep is the swimming pool?
A. 6.25 feet
B. 12 feet
C. 40 feet
D. 100 feet

39) What is the value of x in the following equation?
$$6^x = 7,776$$
A. 3
B. 4
C. 5
D. 6

40) What is the value of x in the following equation?
$$10 + 5(x + 5 - 5x) = 40$$
A. -3
B. $-\frac{1}{4}$
C. $\frac{1}{4}$
D. 3

41) What is the area of the trapezoid?
A. 25
B. 45
C. 100
D. 150

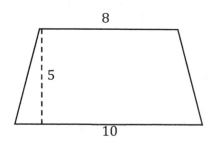

42) 13.125 ÷ 0.005?
   A. 2.625
   B. 26.25
   C. 262.5
   D. 2,625

43) A card is drawn at random from a standard 52–card deck, what is the probability that the card is of Hearts? (The deck includes 13 of each suit clubs, diamonds, hearts, and spades)
   A. $\frac{1}{3}$
   B. $\frac{1}{4}$
   C. $\frac{1}{6}$
   D. $\frac{1}{78}$

44) Ella bought a pair of gloves for $13.59. She gave the clerk $20.00. How much change should she get back?
   A. $4.51
   B. $6.41
   C. $7.51
   D. $8.51

45) $\frac{5 \times 20}{80}$ is closest estimate to?
   A. 1.01
   B. 1.1
   C. 1.3
   D. 1.4

46) If 60% of A is 30% of B, then B is what percent of A?
   A. 2%
   B. 20%
   C. 200%
   D. 300%

47) $\dfrac{3}{4} + \dfrac{\frac{-3}{5}}{\frac{6}{10}} = ?$

A. $\dfrac{1}{4}$

B. $\dfrac{1}{2}$

C. $-\dfrac{1}{4}$

D. $-\dfrac{1}{2}$

STOP

ISEE Middle Level Math Prep 2020-2021

ISEE Middle Level Math

Practice Test 2

2020- 2021

Two Parts

Total number of questions: 84

Part 1 (Quantitative Reasoning): 37 questions

Part 2 (Mathematics Achievement): 47 questions

Total time for two parts: 75 Minutes

ISEE Middle Level Practice Test Answer Sheets

Remove (or photocopy) this answer sheet and use it to complete the practice test.

ISEE Middle Level Practice Test 2

Quantitative Reasoning

1. Ⓐ Ⓑ Ⓒ Ⓓ
2. Ⓐ Ⓑ Ⓒ Ⓓ
3. Ⓐ Ⓑ Ⓒ Ⓓ
4. Ⓐ Ⓑ Ⓒ Ⓓ
5. Ⓐ Ⓑ Ⓒ Ⓓ
6. Ⓐ Ⓑ Ⓒ Ⓓ
7. Ⓐ Ⓑ Ⓒ Ⓓ
8. Ⓐ Ⓑ Ⓒ Ⓓ
9. Ⓐ Ⓑ Ⓒ Ⓓ
10. Ⓐ Ⓑ Ⓒ Ⓓ
11. Ⓐ Ⓑ Ⓒ Ⓓ
12. Ⓐ Ⓑ Ⓒ Ⓓ
13. Ⓐ Ⓑ Ⓒ Ⓓ
14. Ⓐ Ⓑ Ⓒ Ⓓ
15. Ⓐ Ⓑ Ⓒ Ⓓ
16. Ⓐ Ⓑ Ⓒ Ⓓ
17. Ⓐ Ⓑ Ⓒ Ⓓ
18. Ⓐ Ⓑ Ⓒ Ⓓ
19. Ⓐ Ⓑ Ⓒ Ⓓ
20. Ⓐ Ⓑ Ⓒ Ⓓ
21. Ⓐ Ⓑ Ⓒ Ⓓ
22. Ⓐ Ⓑ Ⓒ Ⓓ
23. Ⓐ Ⓑ Ⓒ Ⓓ
24. Ⓐ Ⓑ Ⓒ Ⓓ
25. Ⓐ Ⓑ Ⓒ Ⓓ
26. Ⓐ Ⓑ Ⓒ Ⓓ
27. Ⓐ Ⓑ Ⓒ Ⓓ
28. Ⓐ Ⓑ Ⓒ Ⓓ
29. Ⓐ Ⓑ Ⓒ Ⓓ
30. Ⓐ Ⓑ Ⓒ Ⓓ
31. Ⓐ Ⓑ Ⓒ Ⓓ
32. Ⓐ Ⓑ Ⓒ Ⓓ
33. Ⓐ Ⓑ Ⓒ Ⓓ
34. Ⓐ Ⓑ Ⓒ Ⓓ
35. Ⓐ Ⓑ Ⓒ Ⓓ
36. Ⓐ Ⓑ Ⓒ Ⓓ
37. Ⓐ Ⓑ Ⓒ Ⓓ

Mathematics Achievement

1. Ⓐ Ⓑ Ⓒ Ⓓ
2. Ⓐ Ⓑ Ⓒ Ⓓ
3. Ⓐ Ⓑ Ⓒ Ⓓ
4. Ⓐ Ⓑ Ⓒ Ⓓ
5. Ⓐ Ⓑ Ⓒ Ⓓ
6. Ⓐ Ⓑ Ⓒ Ⓓ
7. Ⓐ Ⓑ Ⓒ Ⓓ
8. Ⓐ Ⓑ Ⓒ Ⓓ
9. Ⓐ Ⓑ Ⓒ Ⓓ
10. Ⓐ Ⓑ Ⓒ Ⓓ
11. Ⓐ Ⓑ Ⓒ Ⓓ
12. Ⓐ Ⓑ Ⓒ Ⓓ
13. Ⓐ Ⓑ Ⓒ Ⓓ
14. Ⓐ Ⓑ Ⓒ Ⓓ
15. Ⓐ Ⓑ Ⓒ Ⓓ
16. Ⓐ Ⓑ Ⓒ Ⓓ
17. Ⓐ Ⓑ Ⓒ Ⓓ
18. Ⓐ Ⓑ Ⓒ Ⓓ
19. Ⓐ Ⓑ Ⓒ Ⓓ
20. Ⓐ Ⓑ Ⓒ Ⓓ
21. Ⓐ Ⓑ Ⓒ Ⓓ
22. Ⓐ Ⓑ Ⓒ Ⓓ
23. Ⓐ Ⓑ Ⓒ Ⓓ
24. Ⓐ Ⓑ Ⓒ Ⓓ
25. Ⓐ Ⓑ Ⓒ Ⓓ
26. Ⓐ Ⓑ Ⓒ Ⓓ
27. Ⓐ Ⓑ Ⓒ Ⓓ
28. Ⓐ Ⓑ Ⓒ Ⓓ
29. Ⓐ Ⓑ Ⓒ Ⓓ
30. Ⓐ Ⓑ Ⓒ Ⓓ
31. Ⓐ Ⓑ Ⓒ Ⓓ
32. Ⓐ Ⓑ Ⓒ Ⓓ
33. Ⓐ Ⓑ Ⓒ Ⓓ
34. Ⓐ Ⓑ Ⓒ Ⓓ
35. Ⓐ Ⓑ Ⓒ Ⓓ
36. Ⓐ Ⓑ Ⓒ Ⓓ
37. Ⓐ Ⓑ Ⓒ Ⓓ
38. Ⓐ Ⓑ Ⓒ Ⓓ
39. Ⓐ Ⓑ Ⓒ Ⓓ
40. Ⓐ Ⓑ Ⓒ Ⓓ
41. Ⓐ Ⓑ Ⓒ Ⓓ
42. Ⓐ Ⓑ Ⓒ Ⓓ
43. Ⓐ Ⓑ Ⓒ Ⓓ
44. Ⓐ Ⓑ Ⓒ Ⓓ
45. Ⓐ Ⓑ Ⓒ Ⓓ
46. Ⓐ Ⓑ Ⓒ Ⓓ
47. Ⓐ Ⓑ Ⓒ Ⓓ

ISEE Middle Level Math Prep 2020-2021

ISEE Middle Level Math
Practice Test 2

Section 1

37 questions

Total time for this section: 35 Minutes

You may NOT use a calculator for this test

1) What is the value of x in the following equation?
$$\frac{7^x}{7} = 343$$

A. 4
B. 5
C. 8
D. 12

2) In following shape y equals to?

A. 120°
B. 30°
C. 25.5°
D. 20°

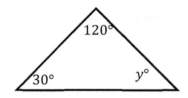

3) Which of the following shows the numbers in increasing order?
A. $\frac{1}{3}, \frac{8}{12}, \frac{4}{7}, \frac{3}{4}$
B. $\frac{1}{3}, \frac{4}{7}, \frac{8}{12}, \frac{3}{4}$
C. $\frac{4}{7}, \frac{3}{4}, \frac{8}{12}, \frac{1}{3}$
D. $\frac{8}{12}, \frac{3}{4}, \frac{4}{7}, \frac{1}{3}$

4) If an object travels at $0.4\ cm$ per second, how many meters does it travel in 5 hours?
A. $88.2\ m$
B. $76.4\ m$
C. $72\ m$
D. $43.2\ m$

5) If the ratio of home fans to visiting fans in a crowd is $3:2$ and all 24,000 seats in a stadium are filled, how many visiting fans are in attendance?
A. 96,000
B. 9,600
C. 960
D. 96

6) What's the approximate circumference of a circle that has a diameter of $17m$?
A. $53.38\ m$
B. $71.9\ m$
C. $97.25\ m$
D. $100\ m$

7) What is the lowest common multiple of 24 and 36?
A. 48
B. 72
C. 108
D. 864

8) What is the area of the shaded region? (one fourth of the circle is shaded) (Diameter = 8)
A. 4π
B. 6π
C. 8π
D. 9π

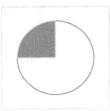

9) An item in the store originally priced at $200 was marked down 30%. What is the final sale price of the item?
A. $240
B. $204
C. $200
D. $140

10) A shirt costing $300 is discounted 15%. After a month, the shirt is discounted another 25%. Which of the following expressions can be used to find the selling price of the shirt?
A. $(300)(0.70)$
B. $(300) - 300(0.30)$
C. $(300)(0.15) - (400)(0.15)$
D. $(300)(0.85)(0.75)$

11) If a car has 70-liter petrol and after one hour driving the car use 5-liter petrol, how much petrol remaining after x-hours?
A. $5x - 70$
B. $70 + 5x$
C. $70 - 5x$
D. $70 - x$

12) Solve for x: $4 + x + 8\left(\frac{x}{4}\right) = 2x + 12$
A. 8
B. 5.5
C. 4
D. 4.5

13) The area of the trapezoid below is 136. What is the value of x?

A. 7
B. 8
C. 10
D. 11

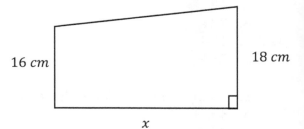

16 cm   18 cm

x

14) Find $\frac{1}{3}$ of $\frac{1}{2}$ of $\frac{3}{5}$ of 280?

A. 28
B. 30
C. 31
D. 2

15) If $x \leq a$ is the solution of $6 + 3x \leq 21$, what is the value of a?

A. $21x$
B. 5
C. -5
D. $15x$

16) 7 liters of water are poured into an aquarium that's $25cm$ long, $5cm$ wide, and $70cm$ high. How many cm will the water level in the aquarium rise due to this added water? ($1\ liter\ of\ water\ =\ 1,000\ cm^3$)

A. 80
B. 56
C. 49
D. 10

17) If $4f + 4g = 4x - 2y$ and $g = 2y - 6x$, what is $2f$?

A. $5x + y$
B. $14x + 3y$
C. $14x - 5y$
D. $y - 3x$

18) What is the value of $\dfrac{-\frac{13}{3} \times \frac{4}{5}}{\frac{10}{30}}$?

A. -10.4
B. 10.4
C. $-\frac{1}{9}$
D. $\frac{1}{9}$

19) What is the perimeter of the following parallelogram?

A. 54

B. 44

C. 24

D. 17

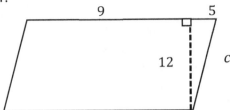

20) In a bundle of 50 fruits, 6 are apples and the rest are bananas. What percent of the bundle is composed of apples?

A. 30%

B. 25%

C. 12%

D. 10%

21) 3 less than twice a positive integer is 71. What is the integer?

A. 37

B. 40

C. 42

D. 44

22) If Joe was making $8.50 per hour and got a raise of $0.35 per hour, approximately what percentage increase was the raise?

A. 2%

B. 2.67%

C. 3.33%

D. 4.00%

23) Which is the equivalent temperature of $140°F$ in Celsius? ($C = Celsius$)

$$C = \frac{5}{9}(F - 32)$$

A. 32

B. 38.5

C. 50

D. 60

24) The average of $14, 16, 21$ and x is 20. What is the value of x?

A. 9

B. 15

C. 18

D. 29

25) What is the value of mode and median in the following set of numbers?

2 ,3, 3, 6, 5, 5, 4, 4, 6, 2, 2

A. Mode: 2 Median:4
B. Mode:2, Median:4
C. Mode:2, 3 Median:5
D. Mode: 3 Median:4

Quantitative Comparisons

Direction: Questions 26 to 37 are Quantitative Comparisons Questions. Using the information provided in each question, compare the quantity in column A to the quantity in Column B. Choose on your answer sheet grid

    A   if the quantity in Column A is greater
    B   if the quantity in Column B is greater
    C   if the two quantities are equal
    D   if the relationship cannot be determined from the information given

26)

Column A	**Column B**
$\dfrac{\sqrt{64-48}}{\sqrt{25-9}}$	$\dfrac{(7-4)}{(8-3)}$

27) $2x^5 - 9 = 477$
$\dfrac{1}{3} - \dfrac{y}{5} = -\dfrac{7}{15}$

<u>Quantity A</u>	<u>Quantity B</u>
x	y

28) The sum of 3 consecutive integers is -45.

Column A	**Column B**
The largest of these integers	-16

29)

Column A	**Column B**
$4^2 - 2^4$	$2^4 - 4^2$

30) A computer costs $250.

Column A	Column B
A sales tax at 8% of the computer cost	$20

31) A

Column A	Column B
The slope of the line $4x + 2y = 7$	The slope of the line that passes through points $(2, 5)$ and $(3, 3)$

32)

Quantity A	Quantity B
The least prime factor of 55	The least prime factor of 210

33)

Column A	Column B
$\sqrt{144 - 81}$	$\sqrt{144} - \sqrt{81}$

34) 6 percent of x is equal to 5 percent of y, where x and y are positive numbers.

Quantity A	Quantity B
x	y

35)

Quantity A	Quantity B
$(-5)^4$	5^4

36)

Quantity A	Quantity B
$(1.88)^4 (1.88)^8$	$(1.88)^{12}$

37) x is a positive number.

Quantity A	Quantity B
x^{10}	x^{20}

ISEE Middle Level Math
Practice Test 2

Section 2

47 questions

Total time for this section: 40 Minutes

You may NOT use a calculator for this test.

1) Which of the following is not synonym for 20^2?
A. 20 cubed
B. 20 squared
C. The square of 20
D. 20 to the second power

2) What is the value of x in the following equation?
$$3^x + 28 = 55$$
A. 3
B. 4
C. 5
D. 6

3) If angles A and B are angles of a parallelogram, what is the sum of the measures of the two angles?
A. 360 degrees
B. 180 degrees
C. 90 degrees
D. Cannot be determined

4) If x = lowest common multiple of 10 and 35, then $\frac{x}{5} + 2$ equal to?
A. 70
B. 58
C. 46
D. 16

5) If the area of the following trapezoid is 30, what is the perimeter of the trapezoid?
A. 25
B. 28
C. 45
D. 55

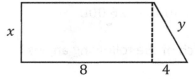

6) A swing moves from one extreme point (point A) to the opposite extreme point (point B) in 20 seconds. How long does it take that the swing moves 5 times from point A to point B and returns to point A?
A. 400 seconds
B. 200 seconds
C. 150 seconds
D. 100 seconds

7) There are 2 cars moving in the same direction on a road. A red car is 12 km ahead of a blue car. If the speed of the red car is 40 $km\ per\ hour$ and the speed of the blue car is $1\frac{2}{5}$ of the red car, how many minutes will it take the blue car to catch the red car?

A. 8.5
B. 15
C. 30
D. 45

8) In two successive years, the population of a town is increased by 10% and 25%. What percent of the population is increased after two years?

A. 25%
B. 35%
C. 36%
D. 37%

9) $5 + 8 \times (-2) - [4 + 22 \times 5] \div 6\ = ?$

A. 120
B. 88
C. −30
D. −20

10) In 1989, the average worker's income increased $2,500 per year starting from $26,000 annual salary. Which equation represents income greater than average?
(I = income, x = number of years after 1989)

A. $I > 2{,}500x + 26{,}000$
B. $I > -2{,}500x + 26{,}000$
C. $I < -2{,}500x + 26{,}000$
D. $I < 2{,}500x - 26{,}000$

11) Which of the following angles is obtuse?

A. 10 degrees
B. 30 degrees
C. 189 degrees
D. 120 degrees

12) Mr. Jones saves $2,500 out of his monthly family income of $65,000. What fractional part of his income does he save?

A. $\frac{1}{26}$

B. $\frac{1}{11}$

C. $\frac{3}{26}$

D. $\frac{2}{15}$

13) Anita's trick–or–treat bag contains 13 pieces of chocolate, 19 suckers, 19 pieces of gum, 25 pieces of licorice. If she randomly pulls a piece of candy from her bag, what is the probability of her pulling out a piece of sucker

A. $\frac{1}{3}$

B. $\frac{1}{4}$

C. $\frac{1}{5}$

D. $\frac{1}{19}$

14) 110 is equal to?

A. $20 - (4 \times 10) + (6 \times 30)$

B. $\left(\frac{11}{8} \times 72\right) + \left(\frac{125}{5}\right)$

C. $\left(\left(\frac{30}{4} + \frac{13}{2}\right) \times 7\right) - \frac{11}{2} + \frac{110}{4}$

D. $(2 \times 10) + (50 \times 1.5) + 15$

15) What is the difference in area between a 8 cm by 4 cm rectangle and a circle with diameter of 8 cm? ($\pi = 3$)

A. 49 cm

B. 40 cm

C. 39 cm

D. 16 cm

16) Solve the following equation?
$$(x^2 + 2x + 1) = 64$$

A. $-9, -7$

B. $-9, 7$

C. -9

D. 7

17) What is ratio of perimeter of figure A to area of figure B?

A. $\frac{1}{3}$

B. $\frac{3}{8}$

C. 3

D. 5

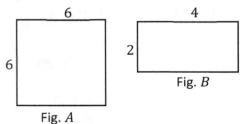

6

4

6

2

Fig. B

Fig. A

18) $\frac{18 \times 21}{5}$ is closest estimate to?

A. 75.6

B. 68.7

C. 50.6

D. 40.7

19) How many possible outfit combinations come from four shirts, two slacks, and five ties?

A. 60

B. 40

C. 16

D. 10

20) When a number is multiplied to itself and added by 9, the result is 25. What is the value of the number?

A. 4 or −4

B. 5 or −5

C. 4

D. 5

21) If you invest \$2,000 at an annual rate of 8%, how much interest will you earn after one year?

A. 16,000

B. 16,00

C. 380

D. 160

22) What is the value of x in the equation: $\frac{x}{4} + \frac{5}{4} = 5$?

A. 15

B. 10

C. 8

D. 5

23) If $y = 5ab + 3b^3$, what is y when $a = 3$ and $b = 2$?

A. 64

B. 65

C. 55

D. 54

24) What is the absolute value of the quantity six minus ten?

A. -4

B. 10

C. -10

D. 4

25) Which of the following angles can represent the three angles of an equilateral triangle?

A. $45°, 90°, 45°$

B. $50°, 50°, 80°$

C. $60°, 60°, 60°$

D. $55°, 35°, 90°$

26) In the following equation, what is the value of $x + y$?

$$9x - 10 = 5\left(\frac{4}{5}x - y\right) + 5$$

A. 15

B. -15

C. 3

D. -3

27) How many tiles of $3\ cm^2$ is needed to cover a floor of dimension $7\ cm$ by $27\ cm$?

A. 12

B. 38

C. 63

D. 66

28) Two-kilogram apple and three-kilograms orange cost $21. If the price of one-kilogram of apple is twice the price of one-kilogram of orange, how much does one-kilogram apple cost?
   A. $8
   B. $6
   C. $4
   D. $1

29) Which is **NOT** a prime number?
   A. 181
   B. 151
   C. 131
   D. 122

30) Each of the x students in a team may invite up to 6 friends to a party. What is the maximum number of students and guests who might attend the party?
   A. $6x + 6$
   B. $6x$
   C. $x + 6$
   D. $7x$

31) Calculate the approximate circumference of the following circle. (the diameter is 10)

   A. 1,267
   B. 314
   C. 31
   D. 10

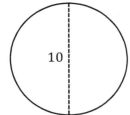

32) 280 $minutes=$...?
   A. 5 $Hours$
   B. 4.6 $Hours$
   C. 3.5 $Hours$
   D. 0.5 $Hours$

33) There are three boxes, a red box, a blue box, and a yellow box. If the weight of the red box is 60 kg and it is 80% of the weight of the blue box, and the weight of the blue box is 120% of the weight of the yellow box, what is the weight of all boxes?
   A. 197.5 kg
   B. 210.5 kg
   C. 280 kg
   D. 320 kg

34) In the figure below, line A is parallel to line B. What is the value of angle x?

A. 35 degree
B. 40 degree
C. 100 degree
D. 140 degree

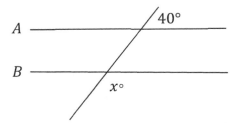

35) Jim drove 350 miles and it took him approximately 9 hours. How many miles per hour was his average speed?
A. about 34.7 miles per hour
B. about 38.8 miles per hour
C. about 48.5 miles per hour
D. about 49.5 miles per hour

36) Three people go to a restaurant. Their bill comes to $58.00. They decided to split the cost. One person pays $7.5, the next person pays 2 times that amount. How much will the third person have to pay?
A. $34.50
B. $35.50
C. $41.00
D. $45.00

37) $\left(\left((-15) + 40\right) \times \frac{1}{5}\right) + (-15)$?
A. 5
B. 10
C. −5
D. −10

38) What is 13,8210 in scientific notation?
A. 138.21×10^3
B. 13.821×10^4
C. 0.13821×10^6
D. 1.3821×10^5

39) What is the value of $(10 − 6)!$?
A. 20
B. 24
C. 26
D. 28

40) What is the perimeter of the below right triangle?

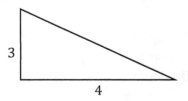

A. 20
B. 18
C. 15
D. 12

41) If 150% of a number is equal to 30% of 80, then what is the number?
A. 16
B. 15.5
C. 14.66
D. 12.25

42) In a department of a company, the ratio of employees with Bachelor's Degree to employees with high school Diploma is 1 to 4. If there are 24 employees with Bachelor's Degree in this department, how many employees with High School Diploma should be moved to other departments to change the ratio of the number of employees with Bachelor's Degree to the number of employees with High School Diploma to 3 to 4 in this department?
A. 64
B. 54
C. 10
D. 12

43) What is x in the following right triangle?

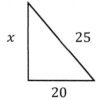

A. $\sqrt{399}$
B. 15
C. 20
D. $\sqrt{402}$

44) The average weight of 20 girls in a class is 65 kg and the average weight of 35 boys in the same class is 72 kg. What is the average weight of all the 55 students in that class?
A. 61.28
B. 61.68
C. 62.90
D. 69.45

45) An angle is equal to one ninth of its supplement. What is the measure of that angle?
A. 18
B. 30
C. 45
D. 60

ISEE Middle Level Math Prep 2020-2021

46) What is the difference of smallest 5–digit number and biggest 5–digit number?
A. 66,666
B. 67,899
C. 88,888
D. 89,999

47) John traveled $140\ km$ in 5 hours and Alice traveled $210\ km$ in 3 hours. What is the ratio of the average speed of John to average speed of Alice?
A. 2 : 5
B. 5 : 2
C. 5 : 9
D. 5 : 6

IF YOU FINISH BEFORE TIME IS CALLED, YOU MAY CHECK YOUR WORK ON THIS SECTION.

STOP

ISEE Middle Level Math Practice Tests Answers and Explanations

Now, it's time to review your results to see where you went wrong and what areas you need to improve!

ISEE Middle Level Math Practice Test 1 Answer Key											
Quantitative Reasoning						**Mathematics Achievement**					
1	B	17	C	33	D	1	C	17	B	33	D
2	B	18	C	34	D	2	C	18	A	34	A
3	B	19	C	35	C	3	A	19	C	35	C
4	A	20	D	36	A	4	D	20	B	36	A
5	D	21	D	37	D	5	D	21	C	37	A
6	C	22	A			6	B	22	B	38	A
7	B	23	A			7	C	23	D	39	C
8	D	24	A			8	C	24	C	40	B
9	A	25	A			9	A	25	D	41	B
10	B	26	A			10	B	26	D	42	D
11	D	27	B			11	C	27	A	43	B
12	D	28	A			12	D	28	D	44	B
13	C	29	D			13	C	29	B	45	C
14	D	30	A			14	D	30	D	46	C
15	D	31	A			15	B	31	D	47	C
16	B	32	A			16	D	32	A		

ISEE Middle Level Math Practice Test 2 Answer Key

Quantitative Reasoning   Mathematics Achievement

1	A	17	C	33	A	1	A	17	C	33	A
2	B	18	A	34	B	2	A	18	A	34	D
3	B	19	A	35	C	3	D	19	B	35	B
4	C	20	C	36	C	4	D	20	A	36	B
5	B	21	A	37	D	5	B	21	D	37	D
6	A	22	D			6	B	22	A	38	D
7	B	23	D			7	D	23	D	39	B
8	A	24	D			8	D	24	D	40	D
9	D	25	B			9	C	25	C	41	A
10	D	26	A			10	A	26	C	42	A
11	C	27	B			11	D	27	C	43	B
12	A	28	A			12	A	28	B	44	D
13	B	29	C			13	B	29	D	45	A
14	A	30	C			14	D	30	D	46	D
15	B	31	C			15	D	31	C	47	A
16	B	32	A			16	B	32	B		

Score Your Test

ISEE scores are broken down by its four sections: Verbal Reasoning, Reading Comprehension, Quantitative Reasoning, and Mathematics Achievement. A sum of the three sections is also reported.

For the Middle Level ISEE, the score range is 760 to 940, the lowest possible score a student can earn is 760 and the highest score is 940 for each section. A student receives 1 point for every correct answer. There is no penalty for wrong or skipped questions.

The total scaled score for a Middle Level ISEE test is the sum of the scores for all sections. A student will also receive a percentile score of between 1-99% that compares that student's test scores with those of other test takers of same grade and gender from the past 3 years.

Use the next table to convert ISEE Middle level raw score to scaled score for application to 7th and 8th grade.

ISEE Middle Level Scaled Scores

Raw Score	Quantitative Reasoning		Mathematics Achievement		Raw Score	Quantitative Reasoning		Mathematics Achievement	
	7th Grade	8th Grade	7th Grade	8th Grade		7th Grade	8th Grade	7th Grade	8th Grade
0	760	760	760	760	26	900	885	885	865
1	770	765	770	765	27	905	890	885	865
2	780	770	780	770	28	910	895	890	870
3	790	775	790	775	29	910	900	890	870
4	800	780	800	780	30	915	905	895	875
5	810	785	810	785	31	920	910	895	875
6	820	790	820	790	32	925	915	900	880
7	825	795	825	795	33	930	920	900	880
8	830	800	830	800	34	930	925	905	885
9	835	805	835	805	35	935	930	905	885
10	840	810	840	810	36	935	935	910	890
11	845	815	845	815	37	940	940	910	890
12	850	820	850	820	38			915	895
13	855	825	855	825	39			920	900
14	860	830	855	830	40			925	905
15	865	835	860	835	41			925	910
16	870	840	860	840	42			930	915
17	875	845	865	840	43			930	920
18	880	845	865	845	44			935	925
19	880	850	870	845	45			935	930
20	885	855	870	850	46			940	935
21	885	860	875	850	47			940	940
22	890	865	875	855					
23	890	870	875	855					
24	895	875	880	860					
25	895	880	880	860					

ISEE Middle LEVEL Math Practice Test 1 Section 1

1) Choice B is correct

$$\frac{-50 \times 0.5}{5} = -\frac{50 \times \frac{1}{2}}{5} = -\frac{\frac{50}{2}}{5} = -\frac{50}{10} = -5$$

2) Choice B is correct

Use the formula for Percent of Change: $\frac{New\ Value - Old\ Value}{Old\ Value} \times 100\%$

$\frac{29-4}{41} \times 100 = -29\%$ (negative sign here means that the new price is less than old price)

3) Choice B is correct

$$562,357,741 \times \frac{1}{10,000} = 56235.7741$$

4) Choice A is correct

30% off equals $24. Let x be the original price of the table. Then:

$$30\%\ of\ x = 24 \rightarrow 0.3x = 24 \rightarrow x = \frac{24}{0.3} = 80$$

5) Choice D is correct

$$3f = 3 \times (3x - 2y) = 9x - 6y, \quad 3f + g = 9x - 6y + x + 5y = 10x - y$$

6) Choice C is correct

$\frac{1}{7} \cong 0.14 \qquad \frac{1}{3} \cong 0.33 \qquad \frac{3}{5} = 0.6 \qquad \frac{3}{4} = 0.75$

7) Choice B is correct

$8^x = 512$, and $512 = 8^3 \rightarrow x = 3$

8) Choice D is correct

Supplementary angles sum up to 180 degrees. x and 35 degrees are supplementary angles. Then: $x = 180° - 35° = 145°$

9) Choice A is correct

If the score of Mia was 40, therefore the score of Ava is 20. Since, the score of Emma was half as that of Ava, therefore, the score of Emma is 10.

10) Choice B is correct

Use the formula of areas of circles. $Area\ of\ a\ circle = \pi r^2 \Rightarrow 49\pi = \pi r^2 \Rightarrow 49 = r^2 \Rightarrow r = 7$, Radius of the circle is 7 Now, use the circumference formula:

146

Circumference $= 2\pi r = 2\pi (7) = 14\pi$

11) Choice D is correct

Let x be the number. Write the equation and solve for x. $\frac{2}{3} \times 15 = \frac{5}{2} \times x \Rightarrow \frac{2 \times 15}{3} = \frac{5x}{2}$, use cross multiplication to solve for x. $2 \times 30 = 5x \times 3 \Rightarrow 60 = 15x \Rightarrow x = 4$

12) Choice D is correct

$1.15 = \frac{115}{100}$ and $8.2 = \frac{82}{10}$ $\rightarrow 1.15 \times 8.2 = \frac{115}{100} \times \frac{82}{10} = \frac{9,430}{1,000} = 9.43 \cong 9.4$

13) Choice C is correct

The perimeter of the trapezoid is 54.

Therefore, the missing side (height) is $= 54 - 17 - 12 - 13 = 12$

Area of the trapezoid: $A = \frac{1}{2} h (b_1 + b_2) = \frac{1}{2} (12) (12 + 13) = 150$

14) Choice D is correct

Add the first 5 numbers. $42 + 46 + 52 + 35 + 58 = 233$

To find the distance traveled in the next 5 hours, multiply the average by number of hours.

$Distance = Average \times Rate = 60 \times 5 = 300$, Add both numbers. $300 + 233 = 533$

15) Choice D is correct
$$average = \frac{10 + 13 + 29 + 37 + 46 + 66 + 100 + 124}{8} = 53.12$$

16) Choice B is correct

$\frac{2}{5}$ of $145 = \frac{2}{5} \times 145 = 58$, $\frac{1}{2}$ of $58 = \frac{1}{2} \times 58 = 29$

17) Choice C is correct

Let x be the original price. If the price of a laptop is decreased by 20% to \$320, then: 80% of $x = 320 \Rightarrow 0.80x = 320 \Rightarrow x = 320 \div 0.80 = 400$

18) Choice C is correct

Let x be the sales profit. Then, 3% of sales profit is $0.03x$. Employee's revenue: $0.03x + 7,500$

19) Choice C is correct

The ratio of boy to girls is $7:4$. Therefore, there are 7 boys out of 11 students. To find the answer, first divide the total number of students by 11, then multiply the result by 7.

$44 \div 11 = 4 \Rightarrow 4 \times 7 = 28$, There are 28 boys and 16 $(44 - 28)$ girls. So, 12 more girls should be enrolled to make the ratio $1:1$

20) Choice D is correct
Let's review the choices:

(A) $\frac{3}{5} > 0.8$ This is not a correct statement. Because $\frac{3}{5} = 0.6$ and it's less than 0.8.

(B) $12\% = \frac{2}{5}$ This is not a correct statement. Because $12\% = 0.12$ and $\frac{2}{5} = 0.4$

(C) $3 < \frac{5}{2}$ This is not a correct statement. Because $\frac{5}{2} = 2.5$ and it's less than 3.

(D) $\frac{7}{6} > 0.8$ This is a correct statement. $\frac{7}{6} = 1.16 \rightarrow 0.8 < \frac{7}{6}$

21) Choice D is correct

Simplify: $6(x + 1) = 4(x - 4) + 20$, $6x + 6 = 4x - 16 + 20$, $6x + 6 = 4x + 4$

Subtract $4x$ from both sides: $2x + 6 = 4$, Add 4 to both sides: $-2 = 2x$, $\quad -1 = x$

22) Choice A is correct

$x = 25 + 135 = 160$

23) Choice A is correct

Petrol of car A in 350 $km = \frac{4 \times 350}{140} = 10$, Petrol of car B in $350 km = \frac{3 \times 350}{140} = 7.5$, $10 - 7.5 = 2.5$

24) Choice A is correct

The diagonal of the square is 6. Let x be the side.

Use Pythagorean Theorem: $a^2 + b^2 = c^2$

$x^2 + x^2 = 6^2 \Rightarrow 2x^2 = 6^2 \Rightarrow 2x^2 = 36 \Rightarrow x^2 = 18 \Rightarrow x = \sqrt{18}$

The area of the square is: $\sqrt{18} \times \sqrt{18} = 18$

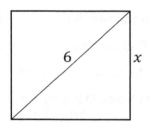

25) Choice A is correct

$-25 - (-66) = -25 + 66 = 66 - 25 = 41$

26) Choice A is correct.

Column A: Simplify. $\sqrt{36} + \sqrt{36} = 6 + 6 = 12$, $\quad$ 12 is greater than $\sqrt{72}$ ($\sqrt{144} = 12$)

27) Choice B is correct.

Column A: The value of x when $y = 12$:

$y = -4x - 8 \rightarrow 12 = -4x - 8 \rightarrow -4x = 20 \rightarrow x = -5$

Column B: -4, -4 is greater than -5.

28) Choice A is correct.

Column A: Use order of operation to calculate the result.

$6 + 4 \times 7 + 8 = 6 + 28 + 8 = 42$

Column B: $4 + 6 \times 7 - 8 \rightarrow 4 + 42 - 8 = 38$

29) Choice D is correct.

Column A: Based on information provided, we cannot find the average age of Joe and Michelle or average age of Michelle and Nicole.

30) Choice A is correct.

Column A: Simplify. $\sqrt{121 - 64} = \sqrt{57}$

Column B: $\sqrt{121} - \sqrt{64} = 11 - 8 = 3$, $\sqrt{57}$ is bigger than 3. ($\sqrt{9} = 3$)

31) Choice A is correct

Volume of a right cylinder $= \pi r^2 h \rightarrow 50\pi = \pi r^2 h = \pi (2)^2 h \rightarrow h = 12.5$, The height of the cylinder is 12.5 inches which is bigger than 10 inches.

32) Choice A is correct.

Simplify quantity B. Quantity B: $(\frac{x}{6})^6 = \frac{x^6}{6^6}$, Since, the two quantities have the same numerator (x^6) and the denominator in quantity B is bigger ($6^6 > 6$), then the quantity A is greater. (remember that x is an integer)

33) Choice D is correct.

Choose different values for x and find the value of quantity A. $x = 1$, then:

Quantity A: $\frac{1}{x} + x = \frac{1}{1} + 1 = 2$,

Quantity B is greater, $x = 0.1$, then:

Quantity A: $\frac{1}{x} + x = \frac{1}{0.1} + 1 = 10 + 1 = 11$,

Quantity A is greater. The relationship cannot be determined from the information given.

34) Choice D is correct

Simply change the fractions to decimals. $\frac{4}{5} = 0.80, \frac{6}{7} = 0.857 ..., \frac{5}{6} = 0.8333 ...,$

As you can see, x lies between 0.80 and 0.857... and it can be 0.81 or 0.84. The first one is less than 0.833... and the second one is greater than 0.833... . The relationship cannot be determined from the information given.

35) Choice C is correct

Choose different values for a and b and find the values of quantity A and quantity B.

$a = 2$ and $b = 3$, then: Quantity A: $|2 - 3| = |-1| = 1$, Quantity B: $|3 - 2| = |1| = 1$

The two quantities are equal. $a = -3$ and $b = 2$, then:

Quantity A: $|-3 - 2| = |-5| = 5$, Quantity B: $|2 - (-3)| = |2 + 3| = 5$

The two quantities are equal. Any other values of a and b give the same answer.

36) Choice A is correct

$$2x^3 + 10 = 64 \rightarrow 2x^3 = 64 - 10 = 54 \rightarrow x^3 = \frac{54}{2} = 27 \rightarrow x = \sqrt[3]{27} = \sqrt[3]{3^3} = 3$$

$$120 - 18y = 84 \rightarrow -18y = 84 - 120 = -36 \rightarrow y = \frac{-36}{-18} = 2$$

37) Choice C is correct.

Quantity A is: $\frac{3+4+x}{3} = 3 \rightarrow x = 2$, Quantity B is: $\frac{2+(2-6)+(2+4)+(2\times2)}{4} = 2$

ISEE Middle LEVEL Math Practice Test 1 Section 2

1) Choice C is correct

50% of 46 is: $\frac{50}{100} \times 46 = \frac{46}{2} = 23$, Let x be the number then: $x = 23 - 5 = 18$

2) Choice C is correct

$4 \times \left(\frac{1}{3} - \frac{1}{6}\right) + 5 = 4 \times \left(\frac{2-1}{6}\right) + 5 = \frac{4}{6} + 5 = \frac{2}{3} + 5 = \frac{17}{3} = 5.6 \ldots$

3) Choice A is correct

20% of $120 = \frac{20}{100} \times 120 = 24$, Let x be the number, then, $x = 24 + 15 = 39$

4) Choice D is correct

Number of pencils are blue$= 85 - 42 = 43$, Percent of blue pencils is: $\frac{43}{85} \times 100 = 50.58\% \cong 51\%$

5) Choice D is correct

$(x + 6)^3 = 64 \rightarrow x + 6 = \sqrt[3]{64} = \sqrt[3]{4^3} = 4 \rightarrow x = 4 - 6 = -2$

6) Choice B is correct

The ratio of red to blue balls is $3:2$. Then: $\frac{3}{2} = \frac{x}{90} \rightarrow x = \frac{90\times3}{2} = 135$

7) Choice C is correct

The perimeter of rectangle is: $2 \times (5 + 8) = 2 \times 13 = 26$

The perimeter of circle is: $2\pi r = 2 \times 3 \times \frac{12}{2} = 36$, Difference in perimeter is: $36 - 26 = 10$

8) Choice C is correct
Let x be the number. Write the equation and solve for x. $(27 - x) \div x = 2$, Multiply both sides by x. $(28 - x) = 2x$, then add x both sides. $27 = 3x$, now divide both sides by 3. $x = 9$

9) Choice A is correct

If $\frac{3x}{2} = 15$, then $3x = 30 \rightarrow x = 10$, $\frac{2x}{5} = \frac{2\times10}{5} = \frac{20}{5} = 4$

10) Choice B is correct

Use this formula: Percent of Change $= \frac{New\ Value - Old\ Value}{Old\ Value} \times 100\%$

$\frac{15,000-20,000}{20,000} \times 100\% = -25\%$ and $\frac{11,200-15,000}{15,000} \times 100\% = -25\%$ (negative signs means the price decreased)

11) Choice C is correct

$\frac{1}{4} = 0.25$ $\frac{7}{9} = 0.77$ $75\% = 0.75$

0.85 is the greatest number provided.

12) Choice D is correct

$$Area = \pi r^2 = \pi \times (\frac{10}{2})^2 = 25\pi = 25 \times 3.14 = 78.5$$

13) Choice C is correct

Only choice C is equal to 90.

$$\left(\left(\frac{3}{2} + 3\right) \times \frac{18}{3}\right) + 63 = \left(\left(\frac{3+6}{2}\right) \times \frac{18}{3}\right) + 63 = \left(\frac{9}{2} \times \frac{18}{3}\right) + 63 = 27 + 63 = 90$$

14) Choice A is correct

All angles in a triangle sum up to 180 degrees. Then:

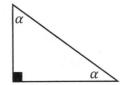

$2\alpha + 90° = 180° \rightarrow 2\alpha = 90 \rightarrow \alpha = 45°$

15) Choice B is correct

First, find the number. Let x be the number. Write the equation and solve for x.

120% of a number is 84, then: $1.2 \times x = 84 \Rightarrow x = 84 \div 1.2 = 70$

90% of 70 is: $0.9 \times 70 = 63$

16) Choice D is correct

Find the difference of each pairs of numbers: $3, 4, 6, 9, 13, 18, 24, \underline{\quad}, 39$

The difference of 3 and 4 is 1, 4 and 6 is 2, 6 and 9 is 3, 9 and 13 is 4, 13 and 18 is 5, 18 and 24 is 6, 24 and next number should be 7. The number is $24 + 7 = 31$

17) Choice B is correct

The sum of angles in rectangle is $360°$

18) Choice A is correct

Let x be the number of shoes the team can purchase. Therefore, the team can purchase $130x$.

The team had $25,000 and spent $15,000. Now the team can spend on new shoes $10,000 at most. Now, write the inequality: $130x + 15,000 \le 25,000$

19) Choice C is correct

The capacity of a red box is 30% greater than a blue box. Let x be the capacity of the blue box. Then: $x + 30\% \ of \ x = 52 \rightarrow 1.3x = 52 \rightarrow x = \frac{52}{1.3} = 40$

20) Choice B is correct

$\frac{1.75}{1.40} = 1.25 = 125\%$

21) Choice C is correct

$$\frac{2}{3} \times 30 = \frac{60}{3} = 20$$

22) Choice B is correct

$$60 \; minutes \; = 1 \; Hours \rightarrow \frac{190}{60} = 3.16 \; Hours$$

23) Choice D is correct

58 is not prime number, it is divisible by 2 and 29.

24) Choice C is correct

The area of the square is 36 inches. Therefore, the side of the square is square root of the area. $\sqrt{36} = 6$ inches, four times the side of the square is the perimeter: $4 \times 6 = 24$ inches

25) Choice D is correct

$13 is what percent of $26? $13 \div 26 = 0.50 = 50\%$

26) Choice D is correct

$\$1.80 \div 36 = \$0.05$

27) Choice A is correct

Let x be one-kilogram orange cost, then: $2x + (2 \times 4.2) = 26.4 \rightarrow 2x + 8.4 = 26.4 \rightarrow$
$2x = 26.4 - 8.4 \rightarrow 2x = 18 \rightarrow x = \frac{18}{2} = \$9$

28) Choice D is correct

Use PEMDAS (order of operation):

$$[6 \times (-24) + 8] - (-4) + [6 \times 5] \div 2 \; = [-144 \; + \; 8] - (-4) + \; [30] \div \; 2$$
$$= \; [-144 \; + \; 8] - (-4) + 15 \; =$$

$$[-136] - (-4) + 15 \; = [-136] + 4 + 15 \; = -117$$

29) Choice B is correct

The width of a rectangle is $3x$ and its length is $5x$. Therefore, the perimeter of the rectangle is $16x$. $Perimeter \; of \; a \; rectangle = 2(width + length) = 2(3x + 5x) = 2(8x) = 16x$

The perimeter of the rectangle is 80. Then: $16x = 80 \rightarrow x = 5$

30) Choice D is correct

The distance between Jason and Joe is 15 miles. Jason running at 5.5 miles per hour and Joe is running at the speed of 7 miles per hour. Therefore, every hour the distance is 1.5 miles less. $15 \div 1.5 = 10$

31) Choice D is correct

$$\Big(((-12) + 20) \times 3\Big) + (-16) = \big((8) \times 3\big) - 16 = 24 - 16 = 8$$

32) Choice A is correct

The percent of girls playing tennis is: $45\% \times 20\% = 0.45 \times 0.20 = 0.09 = 9\%$

33) Choice D is correct

Let x be the original price. If the price of the sofa is decreased by 30% to \$490, then: $70\%\ of\ x = 490 \Rightarrow 0.70x = 490 \Rightarrow x = 490 \div 0.70 = 700$

34) Choice A is correct

There are twice as many girls as boys. Let x be the number of girls in the class. Then:

$x + 2x = 45 \rightarrow 3x = 45 \rightarrow x = 15$

35) Choice C is correct

The ratio of lions to tigers is 5 to 3 at the zoo. Therefore, total number of lions and tigers must be divisible by 8. $5 + 3 = 8$, From the numbers provided, only 97 is not divisible by 8.

36) Choice A is correct

$10x = -55.5 + 15.5 = -40 \rightarrow x = \dfrac{-40}{10} = -4$

37) Choice A is correct

0.25 minutes equals 15 seconds. Then, the number of rotates in 15 second $= \dfrac{400 \times 15}{5} = 1,200$

38) Choice A is correct

Use formula of rectangle prism volume. $V = (length)(width)(height) \Rightarrow$

$$2,500 = (40)\,(10)\,(height) \Rightarrow height = 2,500 \div 400 = 6.25$$

39) Choice C is correct

$$7,776 = 6^5 \rightarrow 6^x = 6^5 \rightarrow x = 5$$

40) Choice B is correct

$10 + 5(x + 5 - 5x) = 10 + 5(-4x + 5) = 40 \rightarrow 10 - 20x + 25 = 40 \rightarrow -20x + 35 = 40$

$$\rightarrow -20x = 5 \rightarrow x = -\dfrac{1}{4}$$

41) Choice B is correct

The area of trapezoid is: $\left(\dfrac{8+10}{2}\right) \times 5 = 45$

42) Choice D is correct

$13.125 \div 0.005 = \dfrac{\dfrac{13,125}{1,000}}{\dfrac{5}{1,000}} = \dfrac{13,125}{5} = 2,625$

43) Choice B is correct

The probability of choosing a Hearts is $\frac{13}{52} = \frac{1}{4}$

44) Choice B is correct

$20 - 13.59 = \$6.41$

45) Choice C is correct

$$\frac{5 \times 20}{80} = \frac{100}{80} = 1.25 \cong 1.3$$

46) Choice C is correct

Write the equation and solve for B: $0.60A = 0.30B$, divide both sides by 0.30, then you will have $\frac{0.60}{0.30}A = B$, therefore: $B = 2A$, and B is 2 times of A or it's 200% of A.

47) Choice C is correct

$$\frac{3}{4} + \frac{\frac{-3}{5}}{\frac{6}{10}} = \frac{3}{4} + \frac{(-3) \times 10}{5 \times 6} = \frac{3}{4} + \frac{-30}{30} = \frac{3}{4} - 1 = \frac{3 - 4}{4} = -\frac{1}{4}$$

ISEE Middle LEVEL Math Practice Test 2 Section 1

1) Choice A is correct

$$343 = 7^3 \rightarrow \frac{7^x}{7} = 7^3 \rightarrow 7^{x-1} = 7^3 \rightarrow x - 1 = 3 \rightarrow x = 4$$

2) Choice B is correct

In triangle sum of all angles equal to $180°$ then: $y = 180° - (120° + 30°) = 180° - 150° = 30°$

3) Choice B is correct

$\frac{1}{3} \cong 0.33$ $\frac{4}{7} \cong 0.57$ $\frac{8}{12} \cong 0.66$ $\frac{3}{4} = 0.75$

4) Choice C is correct

One hour equal to 60 minutes then, $5 \ hours = 5 \times 60 = 300 \ minutes$

One minute equal to 60 seconds then, $300 \ minutes = 300 \times 60 = 18,000 \ seconds$

Distance that travel by object is: $0.4 \times 18,000 = 7,200 \ cm = 72 \ m$

5) Choice B is correct

Number of visiting fans: $\frac{2 \times 24,000}{5} = 9,600$

6) Choice A is correct

Circumference of circle $= 2\pi r = 2\pi \times \frac{17}{2} = 17\pi \sim 53.38\ m$

7) Choice B is correct

The lowest common multiple of 24 and 36 is 72.

8) Choice A is correct

Area of circle with diameter 8 is: $\pi r^2 = \pi \left(\frac{8}{2}\right)^2 = 16\pi$, The area of shaded region is:

$\frac{16\pi}{4} = 4\pi$

9) Choice D is correct

30% of $200 = \frac{30}{100} \times 200 = 60$, Final sale price is: $200 - 60 = \$140$

10) Choice D is correct

To find the discount, multiply the number by $(100\% - rate\ of\ discount)$.

Therefore, for the first discount we get: $(300)(100\% - 15\%) = (300)(0.85)$

For the next 25% discount: $(300)(0.85)(0.75)$

11) Choice C is correct

The amount of petrol consumed after x hours is: $5x$, Petrol remaining: $70 - 5x$

12) Choice A is correct

$4 + x + 8\left(\frac{x}{4}\right) = 2x + 12 \rightarrow 4 + x + 2x = 2x + 12 \rightarrow x = 8$

13) Choice B is correct

The area of trapezoid is: $\left(\frac{16+18}{2}\right)x = 136 \rightarrow 17x = 136 \rightarrow x = 8$

14) Choice A is correct

$\frac{3}{5}$ of $280 = \frac{3}{5} \times 280 = 168$, $\frac{1}{2}$ of $168 = \frac{1}{2} \times 168 = 84$, $\frac{1}{3}$ of $84 = \frac{1}{3} \times 84 = 28$

15) Choice B is correct

$6 + 3x \leq 21 \rightarrow 3x \leq 21 - 6 \rightarrow 3x \leq 15 \rightarrow x \leq \frac{15}{3} \rightarrow x \leq 5$, Then: $a = 5$

16) Choice B is correct

$One\ liter = 1,000\ cm^3 \rightarrow 7\ liters = 7,000\ cm^3$, $7,000 = 25 \times 5 \times h \rightarrow h = \frac{7,000}{125} = 56\ cm$

17) Choice C is correct

$4f + 4g = 4x - 2y \rightarrow 4f + 4(2y - 6x) = 4x - 2y \rightarrow 4f + 8y - 24x = 4x - 2y \rightarrow$

$4f = 28x - 10y \rightarrow 2f = 14x - 5y$

18) Choice A is correct

$$\frac{-\frac{13}{3} \times \frac{4}{5}}{\frac{10}{30}} = -\frac{\frac{13 \times 4}{3 \times 5}}{\frac{10}{30}} = -\frac{\frac{52}{15}}{\frac{10}{30}} = -\frac{52 \times 30}{15 \times 10} = -10.4$$

19) Choice A is correct

Use Pythagorean theorem to find the value of c: $a^2 + b^2 = c^2 \rightarrow 5^2 + 12^2 = c^2 \rightarrow$

$169 = c^2 \rightarrow c = 13$. Perimeter of parallelogram$= (9 + 5 + 13) \times 2 = 54$

20) Choice C is correct

$$\frac{6}{50} \times 100 = \frac{6}{5} \times 10 = 12\%$$

21) Choice A is correct

Let x be the integer. Then: $2x - 3 = 71$, Add 3 both sides: $2x = 74$, Divide both sides by 2: $x = 37$

22) Choice D is correct

$$\frac{0.35}{8.5} \times 100 = 4.11 \approx 4.00$$

23) Choice D is correct

Plug in 140 for F and then solve for C. $C = \frac{5}{9}(F - 32) \Rightarrow C = \frac{5}{9}(140 - 32) \Rightarrow$

$$C = \frac{5}{9}(108) = 60$$

24) Choice D is correct

$$Average = \frac{sum\ of\ terms}{number\ of\ terms} \Rightarrow 20 = \frac{14+16+21+x}{4} \Rightarrow 80 = 51 + x \Rightarrow x = 29$$

25) Choice B is correct

First write the numbers in the order: 2,2,2,3,3,4,4,5,5,6,6

The mode of numbers is: 2   median is: 4

26) Choice A is correct.

Column A: Simplify. $\frac{\sqrt{64-48}}{\sqrt{25-9}} = \frac{\sqrt{16}}{\sqrt{16}} = 1$

Column B: $\frac{(7-4)}{(8-3)} = \frac{3}{5}$

27) Choice B is correct

$2x^5 - 9 = 477 \rightarrow 2x^5 = 477 + 9 = 486 \rightarrow x^5 = \frac{486}{2} = 243 \rightarrow x = \sqrt[5]{243} = \sqrt[5]{3^5} = 3$

$\frac{1}{3} - \frac{y}{5} = -\frac{7}{15} \rightarrow \frac{y}{5} = \frac{1}{3} + \frac{7}{15} = \frac{5+7}{15} = \frac{12}{15} = \frac{4}{5} \rightarrow y = 5 \times \frac{4}{5} = 4$

28) Choice A is correct.

Column A: First, find the integers. Let x be the smallest integer. Then the integers are x, $(x + 1)$, and $(x + 2)$. The sum of the integers is -45. Then:

$$x + x + 1 + x + 2 = -45 \rightarrow 3x + 3 = -45 \rightarrow 3x = -48 \rightarrow x = -16$$

The smallest integer is -16, therefore, the largest integer is bigger than that.

29) Choice C is correct.

Column A: $4^2 - 2^4 = 16 - 16 = 0$

Column B: $2^4 - 4^2 = 16 - 16 = 0$

30) Choice C is correct.

Column A: 8% of the computer cost is 20: $8\% \times 250 = 0.08 \times 250 = 20$

Column B: 20

31) Choice C is correct.

Column A: The slope of the line $4x + 2y = 7$ is -2.

Write the equation in slope intercept form. $4x + 2y = 7 \rightarrow 2y = -4x + 7 \rightarrow y = -2x + \frac{7}{2}$

Column B: The slope of the line that passes through points $(2, 5)$ and $(3, 3)$:

Use slope formula: $slope \ of \ a \ line = \frac{y_2 - y_1}{x_2 - x_1} = \frac{3-5}{3-2} = -2$

32) Choice A is correct

prime factoring of 55 is: 5×11 , prime factoring of 210 is: $2 \times 3 \times 5 \times 7$

Quantity $A = 5$ and Quantity $B = 2$

33) Choice A is correct.

Column A: Simplify. $\sqrt{144 - 81} = \sqrt{63}$

Column B: $\sqrt{144} - \sqrt{81} = 12 - 9 = 3$

$\sqrt{63}$ is bigger than 3. ($\sqrt{9} = 3$)

34) Choice B is correct

6% of x = 5% of $y \rightarrow 0.06 \ x = 0.05 \ y \rightarrow x = \frac{0.05}{0.06} y \rightarrow x = \frac{5}{6} y$, therefore, y is bigger than x.

35) Choice C is correct

Simplify both quantities.

Quantity A: $(-5)^4 = (-5) \times (-5) \times (-5) \times (-5) = 625$

Quantity B: $5 \times 5 \times 5 \times 5 = 625$
The two quantities are equal.

36) Choice C is correct.

Use exponent "product rule": $x^n \times x^m = x^{n+m}$

Quantity A: $(1.88)^4(1.88)^8 = (1.88)^{4+8} = (1.88)^{12}$

Quantity B: $(1.88)^{12}$
The two quantities are equal.

37) Choice D is correct.

Choose different values for x and find the value of quantity A and quantity B.

$x = 1$, then: Quantity A: $x^{10} = 1^{10} = 1$

Quantity B: $x^{20} = 1^{20} = 1$

The two quantities are equal.

$x = 2$, then: Quantity A: $x^{10} = 2^{10}$

Quantity B: $x^{20} = 2^{20}$

Quantity B is greater.

Therefore, the relationship cannot be determined from the information given

ISEE Middle LEVEL Math Practice Test 2 Section 2

1) Choice A is correct

20 cubed is: $20^3 = 8,000$

2) Choice A is correct

$3^x + 28 = 55 \rightarrow 3^x = 55 - 28 = 27$ and $27 = 3^3, 3^x = 3^3 \rightarrow x = 3$

3) Choice D is correct
All angles in a parallelogram sum up to 360 degrees. Since, we only have 2 angles, therefore the answer cannot be determined.

4) Choice D is correct

Prime factorizing of $10 = 2 \times 5$,   Prime factorizing of $35 = 5 \times 7$

$x = LCM = 2 \times 5 \times 7 = 70$, $\frac{70}{5} + 2 = 14 + 2 = 16$

5) Choice B is correct

The area of trapezoid is: $\left(\frac{8+12}{2}\right) \times x = 30 \rightarrow 10x = 30 \rightarrow x = 3$

$y = \sqrt{3^2 + 4^2} = 5$,   Perimeter is: $12 + 3 + 8 + 5 = 28$

6) Choice B is correct

Swing moves once from point A to point B and returns to point A is: $20 + 20 = 40$ seconds

Therefore, for ten times: $5 \times 40 = 200$ seconds

7) Choice D is correct

Speed of the blue car: $1\frac{2}{5} \times 40 = 56$, Difference of the cars' speed: $56 - 40 = 16$, The red car is $12 \ km$ ahead of a blue car. Therefore, it takes the blue car 45 minutes to catch the red car.
$\frac{12}{16} = \frac{3}{3}$ Hour $= 45$ minutes

8) Choice D is correct

The population is increased by 10% and 25%. 10% increase changes the population to 110% of original population. For the second increase, multiply the result by 125%.

$(1.10) \times (1.25) = 1.37 = 137\%$, 37 percent of the population is increased after two years.

9) Choice C is correct

Use PEMDAS (order of operation): $5 + 8 \times (-2) - [4 + 22 \times 5] \div 6 = 5 + 8 \times (-2) - [4 + 110] \div 6 = 5 + 8 \times (-2) - [114] \div 6 = 5 + (-16) - 19 = 5 + (-16) - 19 = -11 - 19 = -30$

10) Choice A is correct

Let x be the number of years. Therefore, $\$2,500$ per year equals $2,500x$. starting from $\$26,000$ annual salary means you should add that amount to $2,500x$. Income more than that is: $I > 2,500 \ x + 26,000$

11) Choice D is correct

Angle between $90°$ and $180°$ is called obtuse angle.

12) Choice A is correct

$2,500$ out of $65,000$ equals to $\frac{2,500}{65,000} = \frac{25}{650} = \frac{1}{26}$

13) Choice B is correct

$Probability = \frac{number \ of \ desired \ outcomes}{number \ of \ total \ outcomes} = \frac{19}{13+19+19+25} = \frac{19}{76} = \frac{1}{4}$

14) Choice D is correct
Only choice D equals 110: $(2 \times 10) + (50 \times 1.5) + 15 = 20 + 75 + 15 = 110$

15) Choice D is correct

The area of rectangle is: $8 \times 4 = 32 \ cm^2$, The area of circle is: $\pi r^2 = \pi \times (\frac{8}{2})^2 = 3 \times 16 = 48$, Difference in area is: $48 - 32 = 16$

16) Choice B is correct
$x^2 + 2x + 1 = (x + 1)^2 \rightarrow (x + 1)^2 = 64 \rightarrow x + 1 = 8 \rightarrow x = 7 \ or \ x + 1 = -8 \rightarrow x = -9$

17) Choice C is correct

Perimeter $A = 4 \times 6 = 24$,   Area $B = 2 \times 4 = 8$, $\frac{24}{8} = 3$

18) Choice A is correct

$$\frac{18 \times 21}{5} = \frac{378}{5} = 75.6$$

19) Choice B is correct

To find the number of possible outfit combinations, multiply number of options for each factor: $4 \times 2 \times 5 = 40$

20) Choice A is correct

Let x be the number, then: $x^2 + 9 = 25 \rightarrow x^2 = 16 \rightarrow x^2 - 16 = 0 \rightarrow (x+4)(x-4) = 0 \rightarrow$ $x = 4 \ or \ x = -4$

21) Choice D is correct

8% of $2,000 = \frac{8}{100} \times 2,000 = \$160$

22) Choice A is correct

$$\frac{x}{4} + \frac{5}{4} = \frac{30}{6} \rightarrow \frac{x}{4} = 5 - \frac{5}{4} = \frac{20-5}{4} = \frac{15}{4} = \frac{x}{4} \rightarrow x = 4 \times \frac{15}{4} = 15$$

23) Choice D is correct

$y = 5ab + 3b^3$, Plug in the values of a and b in the equation: $a = 3$ and $b = 2$

$y = 5 \, (3)(2) + 3 \, (2)^3 = 30 + 3(8) = 30 + 24 = 54$

24) Choice D is correct

$|6 - 10| = |-4| = 4$

25) Choice C is correct

The angles of an equilateral triangle are $60, 60, 60$ degrees.

26) Choice C is correct

$9x - 10 = 5\left(\frac{4}{5}x - y\right) + 5 \rightarrow 9x - 10 = 4x - 5y + 5 \rightarrow 5x - 10 = -5y + 5$

$\rightarrow 5x = -5y + 15 \rightarrow x = -y + 3 \rightarrow x + y = 3$

27) Choice C is correct

The area of the floor is: $7cm \times 27 \, cm = 189 \, cm^2$, The number is tiles needed $= 189 \div 3 = 63$

28) Choice B is correct

Let x be price of one-kilogram of apple and y be price of one-kilogram of orange, then:

$x = 2y, 2x + 3y = 21 \rightarrow 2(2y) + 3y = 21 \rightarrow 7y = 21 \rightarrow y = \frac{21}{7} = 3 \rightarrow x = 2 \times 3 = 6$

29) Choice D is correct

122 is not prime number, it is divided by 2

30) Choice D is correct

Since, each of the x students in a team may invite up to 6 friends, the maximum number of people in the party is 7 times x or $7x$. (one student + 6 friends = 7 people)

31) Choice C is correct

Perimeter= $2\pi r = 2 \times \pi \times \frac{10}{2} = 10 \approx 31.4 \approx 31$

32) Choice B is correct

$60 \ minutes = 1 \ Hour \rightarrow \frac{280}{60} = 4.6 \ Hours$

33) Choice A is correct

Let x be the weight of the red box. Then: Weight of blue box: $0.8x = 60 \rightarrow x = \frac{60}{0.8} = 75$,

Weight of yellow box $1.20y = 75 \rightarrow y = 62.5$

The weight of all boxes: $60 + 75 + 62.5 = 197.5$

34) Choice D is correct

$180° - 40° = 140°$

35) Choice B is correct

Average speed: $\frac{350}{9} = 38.8 \ miles \ per \ hour$

36) Choice B is correct

Let x be the price that third person has to pay then; $58 = 7.5 + (2 \times 7.5) + x \rightarrow$

$x = 58 - 22.5 = 35.5$

37) Choice D is correct

$\left(((-15) + 40) \times \frac{1}{5}\right) + (-15) = \left((25) \times \frac{1}{5}\right) - 15 = 5 - 15 = -10$

38) Choice D is correct

$138,210 = 1.3821 \times 10^5$

39) Choice B is correct

$(10 - 6)! = 4! = 4 \times 3 \times 2 \times 1 = 24$

40) Choice D is correct

$c = \sqrt{4^2 + 3^2} = \sqrt{25} = 5$, Perimeter is: $5 + 3 + 4 = 12$

41) Choice A is correct

Let x be the number then, 150% of $x = 1.5x$, $1.5x = 0.30 \times 80 = 24 \rightarrow x = \frac{24}{1.5} = 16$

42) Choice A is correct

Number of employees with Diploma: $\frac{4 \times 24}{1} = 96$

That means there are 24 employees with Bachelor's Degree and 96 employees with high school diploma. To get the ratio of employees with Bachelor's Degree to the number of employees with High School Diploma 3 to 4, there should be 32 employees with high school diploma. So, 40 $(96 - 32 = 64)$ employees with High School Diploma should be moved to other departments.

43) Choice B is correct

Use Pythagorean theorem: $x^2 + 20^2 = 25^2 \rightarrow x^2 = 25^2 - 20^2 \rightarrow x = \sqrt{25^2 - 20^2} \rightarrow x = \sqrt{625 - 400} = \sqrt{225} = 15$

44) Choice D is correct

$average = \frac{sum\ of\ terms}{number\ of\ terms}$, The sum of the weight of all girls is: $20 \times 65 = 1,300\ kg$

The sum of the weight of all boys is: $35 \times 72 = 2,520\ kg$, The sum of the weight of all students is: $1,300 + 2,520 = 3,820\ kg$, $Average = \frac{3,820}{55} = 69.45$

45) Choice A is correct

The sum of supplement angles is 180. Let x be that angle. Therefore, $x + 9x = 180$, $10x = 180$, divide both sides by 10: $x = 18$

46) Choice D is correct

Smallest 5–digit number is 10,000, and biggest 5–digit number is 99,999. The difference is: 89,999 ,

47) Choice A is correct

The average speed of john is: $140 \div 5 = 28$, The average speed of Alice is: $210 \div 3 = 70$

Write the ratio and simplify. $28 : 70 \Rightarrow 2 : 5$

"Effortless Math Education" Publications

Effortless Math authors' team strives to prepare and publish the best quality ISEE Middle Level Mathematics learning resources to make learning Math easier for all. We hope that our publications help you learn Math in an effective way and prepare for the ISEE Middle Level test.

We all in Effortless Math wish you good luck and successful studies!

Effortless Math Authors

Visit www.EffortlessMath.com
for Online Math Practice

www.EffortlessMath.com

... So Much More Online!

✓ FREE Math lessons

✓ More Math learning books!

✓ Mathematics Worksheets

✓ Online Math Tutors

Need a PDF version of this book?

Visit www.EffortlessMath.com

CPSIA information can be obtained
at www.ICGtesting.com
Printed in the USA
BVHW060952080420
577177BV00008B/135